Management and Industrial Engineering

Series editor

J. Paulo Davim, Aveiro, Portugal

More information about this series at http://www.springer.com/series/11690

Mangey Ram · J. Paulo Davim
Editors

Modeling and Simulation in Industrial Engineering

Springer

Editors
Mangey Ram
Department of Mathematics
Graphic Era University
Dehradun
India

J. Paulo Davim
Department of Mechanical Engineering
University of Aveiro
Aveiro
Portugal

ISSN 2365-0532 ISSN 2365-0540 (electronic)
Management and Industrial Engineering
ISBN 978-3-319-60431-2 ISBN 978-3-319-60432-9 (eBook)
DOI 10.1007/978-3-319-60432-9

Library of Congress Control Number: 2017945709

Printed on acid-free paper

This Springer imprint is published by Springer Nature
The registered company is Springer International Publishing AG
The registered company address is: Gewerbestrasse 11, 6330 Cham, Switzerland

Preface

Modeling and Simulation is arguably one of the most multidimensional topics that can face an industrial engineering nowadays. Modeling and simulation techniques contribute to simplification and economic favoring of the strategy, realization, and tests in industrial engineering. It can also be one of the most important to an organization, regardless of the industry. Reliability, quality, safety, and productivity are all affected by modeling and simulation, whether the issues occur in the office, on the industry floor, or in a warehouse.

The main aim of the book is to give theoretical as well as practical information about modeling and simulation in industrial engineering. This book "Modeling and Simulation in Industrial Engineering" gripped on a comprehensive range of modeling and simulations in industrial engineering areas for different tasks not only in a particular field. Several tools, techniques, and strategies in industrial engineering applications will be discussed in the book.

Dehradun, India — Mangey Ram
Aveiro, Portugal — J. Paulo Davim

Acknowledgements

The Editors acknowledges Springer for this opportunity and professional support. Also, we would like to thank all the chapter authors for their availability for this work.

Contents

Editors and Contributors

About the Editors

Mangey Ram received the BSc degree in science from Chaudhary Charan Singh University, Meerut, India, in 2000, the MSc degree in mathematics from Hemwati Nandan Bahuguna Garhwal University, Srinagar (Garhwal), India, in 2004, and the PhD degree major in mathematics and minor in computer science from G.B. Pant University of Agriculture and Technology, Pantnagar, India, in 2008. He has been a Faculty Member for around nine years and has taught several core courses in pure and applied mathematics at undergraduate, postgraduate, and doctorate levels. He is currently a Professor & Assistant Dean (International Affairs) at Graphic Era University, Dehradun, India. Before joining the Graphic Era University, he was a Deputy Manager (Probationary Officer) with Syndicate Bank for a short period. He is editor in chief of *International Journal of Mathematical, Engineering and Management Sciences*; Executive Editor of *Journal of Graphic Era University*; Associate Executive Editor of *Journal of Reliability and Statistical Studies*; and the Guest Editor and Member of the editorial board of many journals. He is a regular reviewer for international journals, including IEEE, Elsevier, Springer, Emerald, John Wiley, Taylor & Francis, and many other publishers. He has published 101 research publications in IEEE, Springer, Emerald, World Scientific, and many other national and international journals of repute and also presented his works at national and international conferences. His fields of research are reliability theory and applied mathematics. Dr. Ram is a senior member of the IEEE, member of Operational Research Society of India, Society for Reliability Engineering, Quality and Operations Management in India, International Association of Engineers in Hong Kong, and Emerald Literati Network in the UK. He has been a member of the organizing committee of a number of international and national conferences, seminars, and workshops. He has been conferred with "Young Scientist Award" by the Uttarakhand State Council for Science and Technology, Dehradun, in 2009. He has been awarded the "Best Faculty Award" in 2011 and recently Research Excellence Award in 2015 for his significant contribution in academics and research at Graphic Era University.

J. Paulo Davim received the PhD degree in Mechanical Engineering in 1997, the MSc degree in Mechanical Engineering (materials and manufacturing processes) in 1991, the Licentiate degree (5 years) in Mechanical Engineering in 1986, from the University of Porto (FEUP), the Aggregate title (Full Habilitation) from the University of Coimbra in 2005 and the DSc degree from London Metropolitan University in 2013. He is Eur Ing by FEANI-Brussels and Senior Chartered Engineer by the Portuguese Institution of Engineers with a MBA and Specialist title in Engineering and Industrial Management. Currently, he is Professor at the Department of Mechanical Engineering of the University of Aveiro, Portugal. He has more than 30 years of teaching and research experience in Manufacturing, Materials, and Mechanical Engineering with special emphasis in Machining & Tribology. He has also interested in Management & Industrial Engineering and Higher Education for Sustainability & Engineering Education. He has guided large numbers of postdoc, Ph.D. and masters students. He has received several scientific awards. He has worked as evaluator of projects for international research agencies as well as examiner of PhD thesis for many universities. He is the editor in chief of several international journals, guest editor of journals, books editor, book series editor, and scientific advisory for many international journals and conferences. Presently, he is an editorial board member of 30 international journals and acts as reviewer for more than 80 prestigious Web of Science journals. In addition, he has also published as editor (and coeditor) more than 100 books and as author (and coauthor) more than 10 books, 60 book chapters, and 400 articles in journals and conferences (more than 200 articles in journals indexed in Web of Science/h-index 39+ and SCOPUS/h-index 48+).

Contributors

J. Paulo Davim Department of Mechanical Engineering, University of Aveiro, Aveiro, Portugal

Patricio Franco Departamento de Ingenieria de Materiales y Fabricacion, Universidad Politecnica de Cartagena, Cartagena, Spain

Nupur Goyal Department of Mathematics, Graphic Era University, Dehradun, India

Arpan Gupta Indian Institute of Technology, Mandi, India

Aliakbar Montazer Haghighi Department of Mathematics, Prairie View A&M University, Prairie View, USA

Young Jin Han Reliability Engineering Team, Hyosung Corporation, Chang-won, South Korea

Naoto Kaio Department of Economic Informatics, Hiroshima Shudo University, Asaminami-ku, Japan

Ha Won Kim Department of Industrial Engineering, Pusan National University, Busan, South Korea

Amit Kumar Department of Mathematics, Lovely Professional University, Punjab, India

Anuj Kumar Department of Mathematics, University of Petroleum & Energy Studies, Dehradun, India

Li Liu Department of Industrial Engineering, Pusan National University, Busan, South Korea

Pedro J. Lopez-Garcia EXPAL Systems S.A., Madrid, Spain

J. Martínez-Pastor Departamento de Ingenieria de Materiales y Fabricacion, Universidad Politecnica de Cartagena, Cartagena, Spain

Dimitar P. Mishev Department of Mathematics, Prairie View A&M University, Prairie View, USA

Domingo Moratilla EXPAL Systems S.A., Madrid, Spain

Sangeeta Pant Department of Mathematics, University of Petroleum & Energy Studies, Dehradun, India

José Páscoa University of Beira Interior, Covilha, Portugal

Mangey Ram Department of Mathematics, Graphic Era University Dehradun, Uttrakhand, India

Anshumaan Saxena Department of Mechanical Engineering, Graphic Era University, Dehradun, India

Gaurav Sharma Indian Institute of Technology, Mandi, India

Maharshi Subhash Department of Mechanical and Automobile Engineering, Graphic Era University, Dehradun, India

Akshat Tandon Department of Mechanical Engineering, Graphic Era University, Dehradun, India

Michele Trancossi Faculty of Arts, Engineering and Sciences, Sheffield Hallam University, Sheffield, UK

Won Young Yun Department of Industrial Engineering, Pusan National University, Busan, South Korea

Optimization of Forming Processes for Gelled Propellant Manufacturing

J. Martínez-Pastor, Patricio Franco, Domingo Moratilla and Pedro J. Lopez-Garcia

1 Introduction to the Gelled Propellant Manufacturing

The specific methods to be applied for production of gelled propellants depend on the type of propellant (according to their chemical composition and product geometry) and the characteristics of the manufacturing facilities. The processes used for propellant production present similarities with other industrial purposes, but in this case require special procedures, controls and devices. The production equipment and infrastructure applied to produce the propellant grains are complex and specialized. Specific facilities and equipment are necessary for preparing the different compounds of the mixture, executing the mixing process and handling the propellant doughs, as well as for product generation operations such as extruding, laminating, casting or machining, and testing operations for validation and acceptance of intermediate and final energetic products.

One of the main processes of colloidal propellant manufacturing is mixing, either batch mixing or continuous mixing. After testing and acceptance of the individual ingredients, they are usually ground in a mill to obtain the required particle size. All compounds, including the binder, stabilizers, curing agents and burning rate modifiers, are then kneaded in large mixers to form a viscous paste, i.e.

J. Martínez-Pastor · P. Franco (✉)
Departamento de Ingenieria de Materiales y Fabricacion, Universidad Politecnica de Cartagena, C/Doctor Fleming S/N, 30202 Cartagena, Spain
e-mail: patricio.franco@upct.es

J. Martínez-Pastor
e-mail: jm.pastor@upct.es

D. Moratilla · P.J. Lopez-Garcia
EXPAL Systems S.A., Avda del Partenon 16, 28042 Madrid, Spain
e-mail: dmoratilla@expal.biz

P.J. Lopez-Garcia
e-mail: pjlopez@expal.biz

M. Ram and J.P. Davim (eds.), *Modeling and Simulation in Industrial Engineering*, Management and Industrial Engineering, DOI 10.1007/978-3-319-60432-9_1

the propellant doughs. Batch mixers are powerful mixing machines for batch quantities of these very viscous materials. Their purpose is to mix liquids and powders of differing densities into a resultant blend. The bowl holds the ingredients that may be added in sequence, while the rotating blades mix everything together. Batch or continuous mixers can be used to mix accurate quantities of propellant constituents into a very uniform blend. This mixture will burn violently if ignited, and therefore safety procedures become critical. The blend produced is later extruded or cast, and subject to additional process such as curing to create a rubbery material that serves as propellant.

Colloidal propellants are usually generated by extrusion or casting processes. Among these both manufacturing technologies, extrusion is applied for gelled propellants and is generally limited to propellant grains of less than 300 mm in diameter, which have more application to tactical air-to-air, surface-to-air and air-to-surface missiles. Machining of solid propellants is usually carried out by large cutting devices specially modified to accommodate the safety requirements associated with these products. Many of these types of machines are built specifically for a particular propellant grain.

The equipment used for testing and acceptance of the propellant batches by sampling is identical to the equipment applied for checking the chemical and mechanical properties of materials produced for other different applications. In the manufacturing of gelled propellants, the testing devices must allow to perform chemical tests for verifying the desired propellant composition, burning tests for evaluating the burning rate of propellant motors and mechanical tests for ensuring the physical properties required by the designed propulsion system [1].

Gelled propellant doughs exhibit high viscosity and yield stress, and a complex rheological response during their semi-solid state processing, and also present some specific viscoelastic phenomena during their shear flow such as wall slip, migratory effects and extrusion die swell. The characterization of rheological properties of these colloidal compounds for different chemical compositions and manufacturing conditions is a challenging objective in order to optimize those process parameters affecting their viscoelastic response during the forming process, and hence to improve the ballistic quality of the produced propellants [1–3].

In the production systems applied to manufacture gelled propellants with nitrocellulose (NC) as the main active base, the dehydrated nitrocellulose is kneaded along with other additives until the gelation is achieved and the appropriate degree of homogeneity is obtained. Other active bases, such as nitroglycerine (NG) and nitroguanidine (NQ), can be also included in the formulation of colloidal propellants. Subsequently, the resultant gelled propellant doughs are hermetically stored and finally sent to certain forming processes that confer the desired ballistic geometry to the so-called propellant grain [1, 2].

These manufacturing processes, aside from their advantages, present a series of production control limitations that result in the dispersion in the rheological state of the propellant doughs, and thus increase the uncertainty associated with the ballistic response of the final product. Accordingly, the control and minimization of these aspects are essential to comply with the expected ballistic specifications. Greater control of the main manufacturing stages (such as kneading and forming) will provide safer and more efficient processing conditions, since the monitoring and adjustment of process parameters will allow to optimize the rheological state of the propellant doughs to be processed [4–7].

Certain quality control techniques should be used to monitor the main process parameters involved in mixing and extrusion operations, due to their great influence on the ballistic properties of the propellant grain. Using a previously defined self-controlling measuring path, it is possible to identify and quantify certain relationships between process and rheological parameters that can serve to minimize the sources of variability in these manufacturing processes, and so to optimize both the product quality and productivity level [8, 9].

1.1 Manufacturing of Nitrocellulose-Based Colloidal Propellants

The main active base of nitrocellulose-based colloidal propellants is "nitrocellulose" (NC), which is the name that usually receives the nitrate ester of cellulose, whose chemical formulation is $C_{12}H_{14}N_6O_{22}$. This active base is obtained by nitration of cellulose pulps, where the nitrogen content represents the degree of esterification of the resulting nitrating mixture [1, 2].

As can be seen in Fig. 1, nitrocellulose production includes a series of auxiliary processes such as centrifugation, boiling, stabilization and homogenization, through which the adequate dose of nitric and sulphuric acids is adjusted and thus the right level of solubility and viscosity is achieved, conferring to this compound the expected gelation capacity [1].

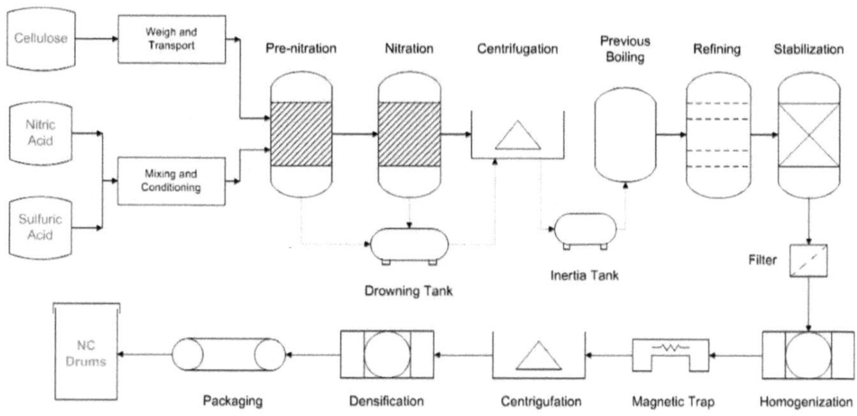

Fig. 1 Basic process flow diagram for nitrocellulose manufacture

The nitroglycerine (NG), whose chemical composition is $C_3H_5N_3O_9$, is produced from high chemical purity glycerine through a continuous process of stirring and cooling until a highly concentrated mixture of nitric and sulphuric acids is obtained, and then the nitroglycerine is extracted in a gravity separator to be subsequently subjected to certain washing processes that eliminate completely the residual dilute acids [1, 2].

The nitrocellulose (NC) and nitroglycerine (NG) generated by means of these processes are the basic raw materials required for the manufacture of double-base gel propellants. When nitroguanidine (NQ) is also added to the mixture, triple-base gel propellants are produced [1, 2].

The manufacturing system for production of colloidal propellants from these raw materials includes a series of processes such us dehydration, kneading (mixing), forming (by extrusion or lamination), cutting, drying, surface treatment and screening (sifting), as illustrated in Fig. 2 [1]. Certain operational details about these industrial processes cannot be explained because they respond to aspects of the productive system that are protected by strict confidentiality clauses. Mixing and extrusion/lamination processes are the main productive stages in terms of their strong influence on the rheological state of the propellant doughs, and consequently the dimensional, mechanical and ballistic properties of the extruded propellant grains.

The highest risk of self-burning is usually considered to be in the extrusion and lamination processes, where propellant doughs (composed of active materials) are subjected to significant pressure levels. However, due to the reduced amount of active material present in each cycle of the extrusion process, the possible human and material damages are of small magnitude. Nevertheless, in certain production stages such as homogenization or cartridge assembly, the potential damages can be

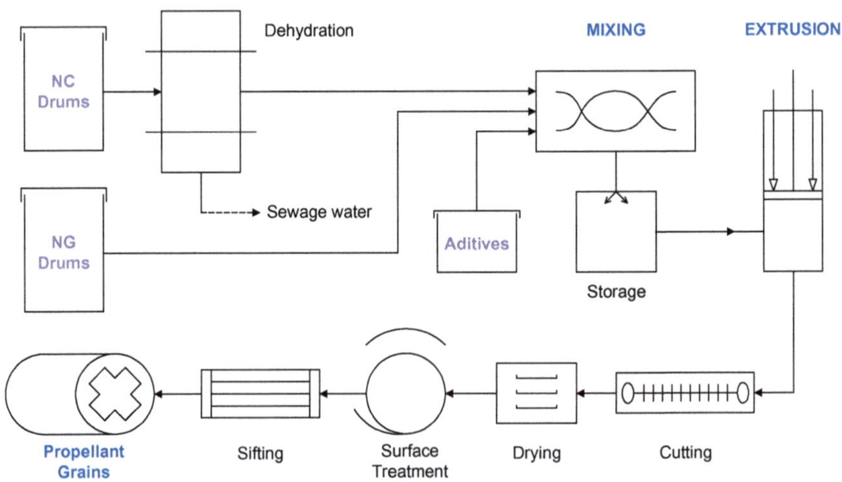

Fig. 2 Basic process flow diagram for the manufacture of colloidal propellants

critical, although the self-burning risk in these processes is remote due to the existence of very strict operational controls and safety conditions.

In those facilities where active materials are processed it is mandatory the use of anti-static work clothes, gloves and footwear, and masks with an appropriate level of filtration, as well as the presence of vapour adsorption and ventilation systems. In general, the major risk factors for workers include ethylic intoxication, vasodilation and headache, produced by contact and inhalation, while the presence of flammable vapours is the main source of risk for the manufacturing facilities, since they can initiate an accident or transmit the effects of fire [1].

1.2 *Manufacturing Processes Associated with the Rheological Behaviour of Gelled Propellants*

In this section, the fundamentals of the mixing and forming (extrusion or lamination) stages will be described, as crucial stages for the rheological behaviour of the propellant doughs during their processing, and hence for the expected quality of final propellant grains.

1.2.1 Batch Mixing of Nitrocellulose Gels

Batch mixing is used when kneading and extrusion processes are carried out in different machines during the manufacturing of gelled propellants. This section is focused on this mixing technology. Similar explanations could be provided in the case of continuous mixing.

Mixing is an essential technology in production of nitrocellulose gels, in order to obtain propellant doughs with the appropriate mechanical and physical properties for other manufacturing stages such as extrusion, lamination and cutting [1]. During this kneading process, not only the merging of the solid and fluid compounds contained in the chemical composition of each specific colloidal propellant, but also the expected chemical reactions for the gelation of the initial mixture are originated, which needs a certain activation or formation energy until its transformation into a concentrated suspension, and then the adequate rheological state is obtained, and the gelation and homogenization are reached [3, 4].

Due to the intrinsic variability in the properties of raw materials and the complex phenomena that take place in this process, the mixing stage usually confers a certain level of dispersion to the rheological behaviour of the resultant propellant doughs, and so it must be conveniently optimized and controlled to avoid an excessive inaccuracy in mechanical, geometrical and ballistic properties of final product [3, 10]. For this reason, the rheological properties of propellants doughs are one of the main aspects that are commonly employed to evaluate the performance of batch

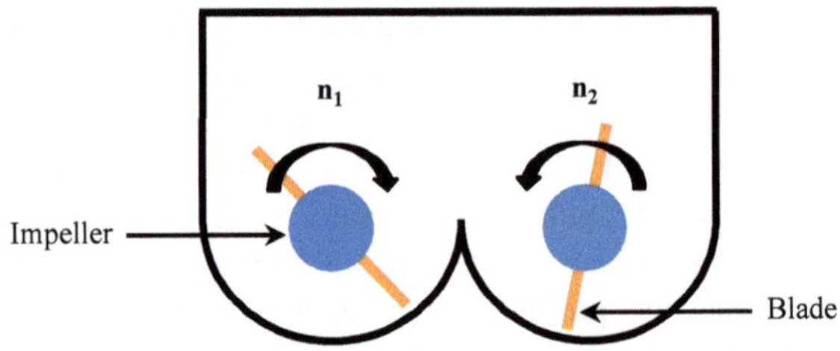

Fig. 3 Cross-sectional schematic of a batch mixer with two co-rotating impellers of different rotational speeds

mixing process, whose parameters strongly affect the viscoelastic state of the kneaded compounds [11].

During the batch mixing process, the nitrocellulose and nitroglycerine are introduced in a batch mixer with the rest of additives, according to a particular mixing procedure that defines the sequences, quantities and times for the addition of each compound according to the chemical formulation of each commercial propellant, allowing the gelation of the product and providing specific propellant doughs to be extruded or laminated [1]. The sealing of the machine framework should be guaranteed, in order to prevent the loss of volatile solvents during the kneading process.

The batch mixers often have two impellers that co-rotate in opposite directions and at different speeds, and distinct configurations can be selected for their blades (Fig. 3). More information about the configuration of this type of batch mixers can be found in specialized publications such as the Ref. [4].

Once the kneading process is completed, the propellant doughs are manually downloaded from the batch mixer and stored in anti-static and hermetic bags, preventing the degradation of the product as a result of the high volatility of their additives [12]. Lately, propellants doughs are sent on demand to the ram extrusion press, where this material to be processed is also loaded manually.

1.2.2 Ram Extrusion of Gelled Propellant Doughs

The extrusion process is responsible for conferring to the propellant grains the cross-sectional geometry required by ballistic specifications, i.e. the number, distribution and size of holes in the propellant grain.

Two types of extrusion machines can be found in the industry: ram extruders and screw extruders. This work is focused on the extrusion of colloidal propellants by ram extruders, which consist of the use of a ram press to carry out the extrusion process. In ram extrusion, the gelled material (propellant doughs) is displaced to the extrusion die by the action of the ram inside the cylinder of the extrusion press, and then the final product takes the cross-sectional geometry defined by the holes of extrusion die (Fig. 4) [1].

Screw extruders can be also applied in the manufacturing of colloidal propellants. These machines are constituted by one or two screws (for single-screw or twin-screw extrusion, respectively), which knead the compounds of the propellant

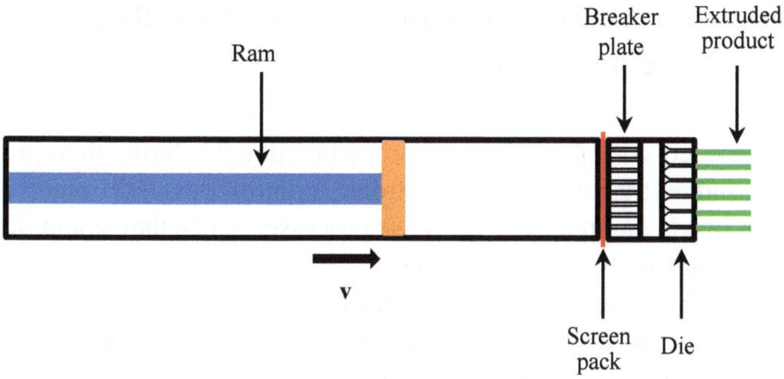

Fig. 4 Longitudinal section schematic of a ram extruder for manufacturing of colloidal propellants

formulation and also transport rotationally the gelled material to the extrusion die, located at the end of the axis of the screw extrusion machine [13].

Both the ram extruders and screw extruders can be adapted for processing distinct types of colloidal propellants with different cross-sectional configurations, simply by changing the extrusion die to be used in this forming process. The number, location and cross-sectional geometry of the channels of the extrusion die are properly designed according to the expected ballistic properties for each type of propellant [13, 14].

Before its access to the extrusion die, the material must pass through a filtering system, composed by the screen filters and a breaker plate (Fig. 4). Screen filters are used for retaining undesirable external particles that could be present inside the propellant doughs, coming from its manipulation in the previous manufacturing stages. The number of screens and the mesh size for each one of them are usually determined by the diameter of the breaker plate's boreholes, throughput rate to be considered, characteristic diameter of the foreign particles and apparent viscosity of the material [13].

The breaker plate has a triple function in the extrusion process: firstly to hold the screen filters, secondly to provide an adequate sealing between the cylinder and the extrusion die (so it must be sufficiently thick to act as a mechanical sealing), and finally to homogenize the viscous or viscoplastic state of the material before their processing by the extrusion die [13].

The extrusion system can present some differences according to the configuration of the extrusion process, the required layout distribution in the factory, the characteristics of propellants to be manufactured and the number of product outputs for each extrusion press.

As essential auxiliary devices that are common to the majority of extrusion process facilities, the cooling circuit, the cutting system and the puller can be mentioned. More detailed information can be found in specialized publications such as the reference [13].

2 Process and Material Parameters for Propellant Production Controlling

Due to the elevated self-burning risk of propellant doughs during their processing at the mixing and extrusion stages, electronic control devices cannot always be implemented inside the machines for these processes, since they could increase this risk or they could imply an excessive cost.

The sensors associated with such control devices should comply with ATEX normative, in order to avoid that they could originate static electricity, increasing the self-burning risk of this equipment. The requirement of using ATEX devices supposes a significant increase in the cost of these control systems, and their installation implicates to carry out certain modifications in the machinery.

2.1 Process Parameters to Control During the Propellant Production

The process control can be made by electronic control devices or by specific procedures for production monitoring. Alternative external tools for production monitoring can be applied in order to avoid machinery modifications and excessive control costs. In this section, the basic parameters of mixing and extrusion processes, and the external control tools for controlling these production stages are described.

2.1.1 Process Parameters in Batch Mixing

Before to describe the procedure to be applied for controlling or monitoring the batch mixing process, the process parameters to be subject to the process control system must be properly identified.

One of the main parameters of batch mixing process is the mixing power $P_m(t)$ (or mixing energy per unit of time), which can be expressed in terms of the rotational speed of the impellers n and the instant torque T(t) developed by the batch mixer:

$$P_{\mathrm{m}}(t) = 2\pi n T(t) \tag{1}$$

where n is the rotational speed of the driven shaft (i.e. the faster rotor when having more than one impeller) and T is the instant torque exerted during the mixing process [4, 5].

From this equation, the specific energy that must be invested during the mixing process (E_{m}) for a given amount of product to be kneaded (m) can be expressed by the following equation [5, 19]:

$$E_{\mathrm{m}} = \frac{2\pi n}{m} \int_{0}^{t_{\mathrm{m}}} T(t)\mathrm{d}t \tag{2}$$

where t_m is the total time of the kneading process that will be required to obtain propellant doughs with the expected viscoelastic state. This equation serves to monitor the mixing process.

The mixing power of this expression corresponds to the electrical power consumption by the batch mixer during these kneading processes. If an electronic control system is not contained in the mixing machine, the measuring of this parameter can be made by a power monitoring device to be properly installed for this purpose in the electrical panel of the extrusion machine.

The temperature registered during the mixing stage is also assumed as a basic process parameter for an adequate production controlling, in order to make possible the identification of the rheological properties and homogeneity level to be achieved as a result of this process. For this purpose, the room temperature at the beginning and at the end of the mixing process is considered as process parameters for the proposed methodology.

As consequence, for an adequate monitoring of the batch mixing stage, the following four basic process parameters must be controlled (considering a fixed rotational speed of the driven shaft):

- Mixing time t_m
- Mixing energy E_m
- Product weight m
- Mixing temperature $T_m(t)$

This section is limited to identifying the parameters that must be monitored during the mixing stage of propellant production, and the specific procedures that must be applied to guarantee the optimized performance of this manufacturing process will be described in later sections of this work.

2.1.2 Process Parameters in Ram Extrusion

In ram extrusion processes, the following equation can be used to determine the instantaneous power $P_e(t)$ applied by the extrusion press, according to the extrusion pressure and ram velocity [13]:

$$P_{\mathrm{e}}(t) = \pi r_0^2 p_{\mathrm{e}}(t) v_{\mathrm{e}}(t) \tag{3}$$

where $p_e(t)$ is the steady extrusion pressure, $v_e(t)$ is the ram velocity, and r_0 is the radius of the cylinder of the extrusion press.

The extrusion pressure and ram velocity are the basic parameters to optimize the performance of the extrusion process, since the flow conditions of the propellant doughs will be conditioned by the values adopted for these process parameters. These process parameters are subject to severe safety criteria in order to prevent the elevated self-burning risk that exists in these forming processes, as a consequence of the active materials contained in the chemical composition of propellants.

The steady extrusion pressure and extrusion ram velocity can be identified in the control panel of the ram press, which usually provides the instantaneous values of these process parameters during the propellant manufacturing.

In addition to these parameters, the extrusion temperature $T_e(t)$ must be also considered for controlling or monitoring the extrusion stage. The thermal control of this forming process is recommended because the temperature level can affect the viscosity of propellant doughs and the surface quality of extruded products [15].

An adequate range for extrusion temperature helps to satisfy the expected surface accuracy in the final product, while higher temperatures can lead to the formation of irregular surfaces and lower temperatures favour the appearance of streaks on the surface of extruded grains. When the temperature inside the cylinder is not enough to soften the entire volume of the material, protuberances can be originated in the propellant grains, and when the temperature is excessively high, the material tends to be difficult to handle due to its low viscosity [16, 17].

The objective of the temperature control is to ensure a uniform heating of the material inside the cylinder, and thus to avoid possible discontinuities in the material flow as a result of uneven heating in certain areas. The ideal operating temperature depends on the type of material flow, the production rate and the heating system.

The established extrusion ram velocity affects significantly the temperature level to be imposed. For higher extrusion velocities, it could be necessary to increase the extrusion temperature, depending on the thermal conductivity of the propellant dough. Further information about the control of extrusion temperature can be found in specific publications such as Refs. [7, 10, 11, 15–17].

During the production of energetic materials, frequently it is not applied a thermal control integrated in the control system of the extrusion press, and so the measurement of material temperature must be carried out manually just before and after the extrusion process.

Finally, there is an additional parameter to be considered during the extrusion process when an intermediate storage stage is applied after the mixing process. In this case, the kneaded gelled propellant doughs are stored in hermetical bags during a certain time interval, depending on the extrusion press availability, and the storage time t_s must be also controlled due to its great influence of the rheological properties of the material.

As a summary, the essential process parameters that must be monitored to implement an adequate production control during the extrusion stage of propellant manufacturing are the following ones:

- Extrusion pressure $p_e(t)$
- Ram velocity $v_e(t)$
- Extrusion temperature $T_e(t)$
- Storage time t_s (if applicable)

2.2 Material Parameters to Control During the Propellant Production

During the production of gelled propellants, certain material parameters must be identified and measured to guarantee the adequate process performance, considering the relationship among these variables and the main process parameters. These material parameters are usually divided in three categories: rheological, elastic and thermodynamic properties [14].

There is also other kind of material properties, such as average particle size, particle size distribution, packing fraction, molecular weight (MW) or molecular weight distribution (MWD), which must be verified for each production batch in order to assure the expected quality of final product. Dynamic mechanical analysis (DMA), scanning electron microscopy (SEM), thermo gravimetric analysis (TGA), X-ray diffraction (XRD) or other similar techniques can be applied to carry out the testing of these variables [5, 6, 18].

However, when the aim of the analysis is to establish a simple and robust process control for the propellant manufacturing system, the controlling and monitoring routines can be specially limited to the rheological and physical parameters highlighted in the following sections. The elastic behaviour of the material can be also basically described by the oscillatory rheological moduli that will be explained bellow, since they serve to identify the viscous and elastic response of the analysed compounds.

2.2.1 Rheological Parameters

Polymeric and elastomeric materials exhibit both viscous and elastic behaviour, and so their rheological response is between ideal fluids and Hookean solids. In addition, concentrated suspensions such as the propellant doughs are characterized by a non-Newtonian response under shear flow (which is the case, e.g., of material flow during the extrusion process), and thus the determination of their rheological properties becomes more complicated [19–23].

The rheological behaviour of a certain material is usually expressed by a material law (or viscous law), which is an equation that establishes the relation between the velocity and stress fields during the shear flow of this compound. This equation is usually focused on describing the viscoplastic response of the analysed material, but misses its elastic properties. This material law represents the rheological properties by means of viscous, viscoplastic or pseudoplastic models such as the models developed by Cross, Carreau, Bingham, Herchel–Bulkley, Casson and other different authors. There are also other models that include the viscoelastic behaviour, for example the models proposed by Maxwell, Phan–Thien–Tanner (PTT), Giesekus, Leonov and Criminale–Ericksen–Filbey (CEF), which are valid for distinct shear flow conditions like the extrusion flow [24–27].

In any case, when the objective is to know the relationship between the process and material parameters in order to enhance the production performance, an exhaustive evaluation of the complete rheological behaviour of the material is not required and, by the contrary, the measuring of certain rheological parameters representative of the material viscoelastic state is recommended. Such variables can be registered by small-amplitude oscillatory shear (SAOS) tests, which provide the oscillatory rheological moduli that serve to describe the viscoelastic state of the material within a wide range of strain rates [24–26].

Using a rotational rheometer with a testing configuration of parallel plates, the oscillatory rheological tests can impose over the samples an oscillatory shear stress within certain range of amplitudes (for a given deformation time or frequency), or an oscillatory shear stress of constant amplitude within a certain range of frequencies. The rotational rheometers are limited to small sample volumes in comparison with capillary rheometers, and nonlinear elastic properties, elastic instabilities and die swell (expressed as an effect of normal stresses) associated with the flow of concentrated suspensions cannot be directly measured by SAOS tests [23].

Nevertheless, the measuring of the oscillatory rheological moduli is enough for a process quality control that includes the monitoring of the viscoelastic behaviour of propellant doughs. The storage modulus G' represents the energy conserved through elasticity during a deformation cycle (solid-like response), while the loss modulusG'' refers to the energy dissipated by viscous heating (viscous-like response). These rheological properties are evaluated in the range characterized by a linear viscoelastic behaviour, which is the linear viscoelastic region (LVR). Within the LVR, the rheological moduli G' and G'' are independent of the oscillatory amplitude imposed for the strain rate or shear stress for a given frequency [24, 25].

During SAOS analysis, the rotational tests impose small deformations in which the shear stress τ is described by a sin (or cosine) wave and presents a certain phase gap δ with regard to the sin (or cosine) wave associated with the strain rate $\dot{\gamma}$, as illustrated in Fig. 5. The shear stress and strain rate extracted from SAOS tests can be expressed by the following equations [24–26]:

$$\tau(\omega t) = \tau_0 \sin(\omega t) \tag{4}$$

$$\dot{\gamma}(\omega t) = \dot{\gamma}_0 \sin(\omega t + \delta) \tag{5}$$

where ω is the angular speed of the rotating plate, and τ_0 and γ_0 are the amplitudes of the sinusoidal signals for shear stress and strain rate, respectively.

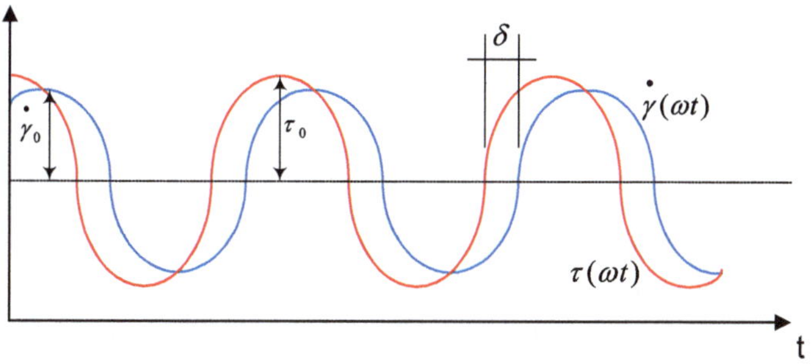

Fig. 5 Graphic representation of sinusoidal signals of shear stress and strain rate during SAOS tests

Decomposing the oscillatory stress signal in two different waves, including a term in phase with the strain rate wave and another term not in phase with it, the following expression in complex notation can be derived:

$$\tau(\omega t) = {}_0 e^{i\omega t} \rightarrow \tau(\omega t) = \tau_0' \sin(\omega t) + \tau_0'' \cos(\omega t) \tag{6}$$

This equation can be also expressed in terms of the storage modulus G' and loss modulus G'' as follows:

$$\tau(\omega t) = G'\gamma_0 \sin(\omega t) + G''\gamma_0 \cos(\omega t) \tag{7}$$

The amplitude of sinusoidal signals for shear stress and strain rate can be related between them from the complex modulus G^*:

$$\tau_0 = |G^*|\gamma_0 \tag{8}$$

$$G^* = G' + iG'' \rightarrow G^* = \sqrt{G'^2 + G''^2} \tag{9}$$

And finally, the storage modulus G' and loss modulus G'' can be expressed as follows as a function of the viscosity, according to the definition of complex viscosity η^*:

$$\eta^* = -i\omega G^* = \eta' - i\eta'' \tag{10}$$

$$G' = \eta'' \cdot \omega \tag{11}$$

$$G'' = \eta' \cdot \omega \tag{12}$$

where η' is the dynamic viscosity and η'' represents the elasticity contained in the complex viscosity.

In addition to the referred to SAOS tests, the rotational rheology provides other numerous testing techniques that also can help to explain in depth the behaviour of each propellant dough formulation, although they were not included inside the proposed production controlling methodology. These other techniques comprise, for instance, stress relaxation, creep, creep-recovery and torsional flow testing, some of which shall be applied to obtain the totality of rheological data required for numerical modelling purposes [28, 29], and to analyse the material microstructure and other properties related to the processability of these special semi-solid compounds.

It is important to clarify that the rheological parameters δ, η^*, η' and η'' do not provide additional information in comparison to the rheological moduli G' and G''. Hence, the unique rheological parameters that are included in the methodology described in this work for controlling and monitoring of propellant manufacturing are the storage modulus and loss modulus.

Nevertheless, once the storage modulus G' and loss modulus G'' for each propellant formulation are identified, the complex modulus G^* will be assumed as control parameter instead of these storage moduli in order to simplify the process monitoring and the possible correlation with the other material and process variables. The complex modulus can be easily deduced from the storage modulus G' and loss modulus G'' according to Eq. (9):

- Storage modulus G'
- Loss modulus G''
- Complex modulus G^*

2.2.2 Physical Parameters

For an adequate analysis of the mixing and extrusion stages during propellant production, the basic physical properties to be registered include the density of the material (ρ), the heat conductivity (λ_T) and the heat capacity at constant pressure (C_p). The measuring of the heat conductivity and heat capacity at constant pressure is needed in the case of non-isothermal extrusion processes, in order to represent the heat transfer by the process control system.

Sometimes these physical properties or the material depend not only on the temperature, but also on the pressure applied during the extrusion stage. Nevertheless, the influence of the pressure on the physical and thermodynamic properties of the material is usually negligible within the typical range of operative pressures, generally lower than 200 bar in these extrusion processes. On the other hand, there are some assumptions that could be applied when the effect of pressure should be considered. In general, these physical properties can be expressed exclusively as a function of temperature during the manufacture of gelled propellants [14].

To simplify the proposed methodology for production controlling and monitoring, the physical parameters to be registered can be limited to the density of the material to be processed. It is enough in cold extrusion processes and when non-isothermal analysis is not required:

- Density ρ.

3 Self-controlling Measuring Path for Propellant Production

Once the representative process and material parameters for the quality control of the mixing and extrusion stages were properly identified, the routines to be carried out for assuring the optimized values for these variables during the production of nitrocellulose gelled propellants must be established.

For each manufacturing process under analysis (i.e. the batch mixing and ram extrusion of propellant doughs), the distinct material and process parameters must be assumed as input or output variables, and certain variables will serve to guarantee the optimized performance of the production system. For example, the initial mixing temperature $T_{i,m}$ and effective mixing time t_m shall be considered as input parameters for the batch mixing stage, while the mixing energy E_m will allow to obtain the desired degree of gelation in the resulting compounds, which will be described by the rheological moduli G' and G'' [5, 15].

The main challenge for the implementation of this quality control methodology consists of identifying the optimized reference values that should be adopted for each input and output parameters, and the possible variations to be introduced in the process parameters as a response to the initial material parameters in order to achieve the expected properties in the extruded propellant grains. In addition, this methodology will serve to clearly specify the material and process parameters that must be examined during the propellant production and the controls to be executed for this purpose.

The methodology proposed for an enhanced production controlling and monitoring includes the use of a certain measuring path to assure the validity of each production batch. The control parameters to be checked during the execution of batch mixing and ram extrusion of nitrocellulose gels are represented in Fig. 6, as well as the self-controlling measuring path to be followed.

These variables include material properties such as the density and rheological moduli of propellant doughs at the end of the mixing stage and before the extrusion stage, and the main process conditions of batch mixing, intermediate storage and ram extrusion, i.e. the initial and final temperatures associated with the mixing and extrusion processes, effective mixing time, overall mixing energy, product weigh to be considered in the mixing stage, storage time, steady extrusion pressure and extrusion ram velocity.

The self-controlling measuring path to be followed is also illustrated with higher details in Fig. 7, including the successive steps that make possible the measuring of the distinct control parameters.

A process database must be developed to collect the historical data about the mixing and extrusion processes for each commercial product (i.e. for each specific propellant formulation). This database can serve to establish the optimum values of the control parameters when testing new formulations or variations of a certain product, as well as to identify potential productive failures and to develop preventive actions for correcting these problems.

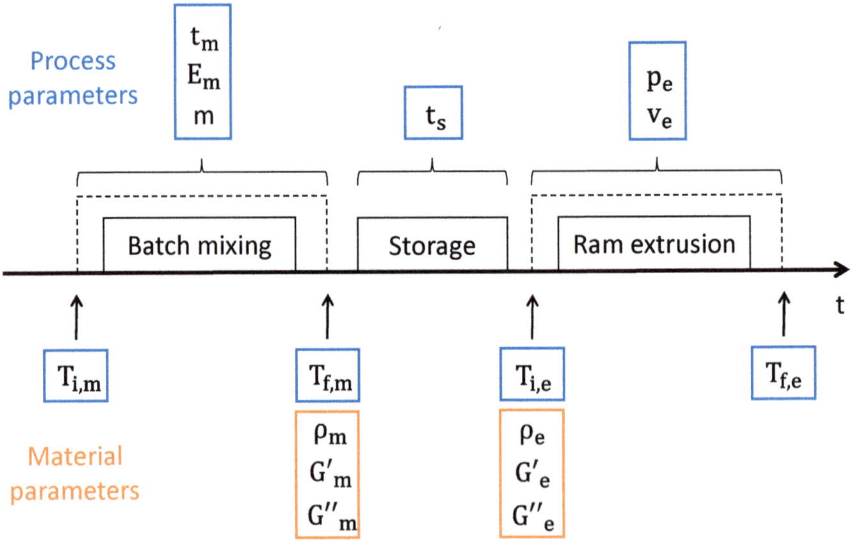

Fig. 6 General diagram about the self-controlling measuring path for mixing and extrusion processes during colloidal propellant production

The data registered according to this measuring path can help to deduce quantitative relations between the material properties and processing conditions, and so the main parameters of these manufacturing stages could be adjusted or regulated in order to provide the desired rheological behaviour in the propellant doughs and the optimum process performance.

Scatter diagrams can be employed to found the empirical correlations that connect the above-mentioned process and material parameters, and also to discuss the nature of their possible influence on the expected productivity level and quality of the extruded propellant grains.

4 Measuring Devices for Process and Material Parameters

In order to attend the self-controlling measuring path proposed in this methodology, the following testing devices should be used for an appropriate quality control of batch mixing and ram extrusion processes during colloidal propellant manufacturing:

- Power monitoring device: This electronic device shall be connected to the power line that feeds the electrical engine of each mixing machine, in order to register the main parameters of mixing process by indirect measuring. By means of this equipment, the instantaneous electrical power consumption can be registered as an index that serves to identify the torque and energy applied to make possible

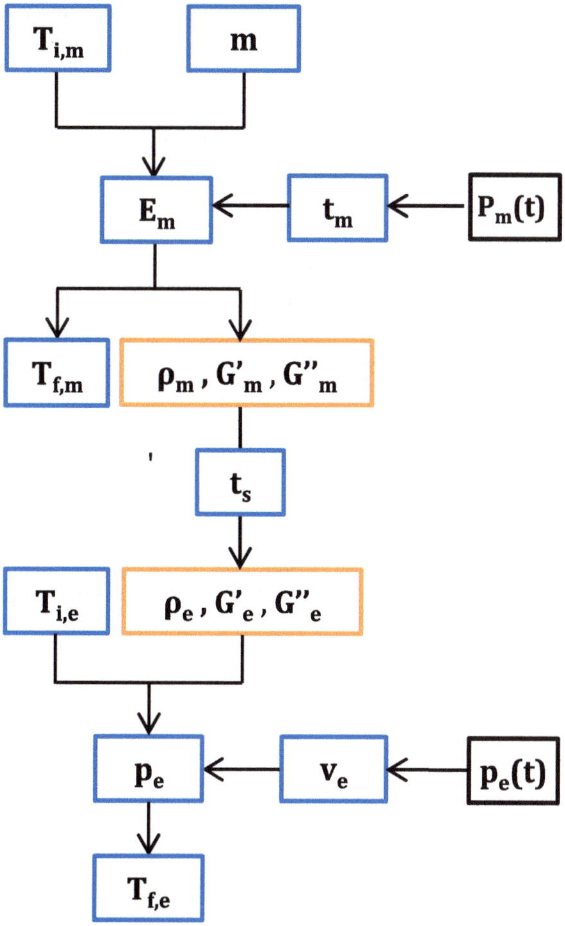

Fig. 7 Detailed diagram about the self-controlling measuring path among the control parameters of batch mixing and ram extrusion

the rotational speed of the batch mixer shaft. The resultant curve depicting the electrical power variation during the kneading process will depend on the specific formulation of each gelled propellant, and the complex material transformations that take place during this process, and will serve to assure the validity of energetic materials obtained by each production batch. The overall mixing energy E_m required to complete the gelation and homogenization of propellant doughs for each specific formulation can be calculated from the instantaneous electrical power consumption $P_m(t)$ and the effective mixing time t_m.

- Pyrometer: Thermometer that must comply with ATEX normative, which consists of the regulations applied for equipment in explosive atmospheres. It includes an immersion thermocouple type-K in order to measure the material temperature. By introducing the thermocouple inside the propellant dough, the temperature of these semi-solid materials can be registered easily and without any risk. This kind of thermometers serves to evaluate the temperatures $T_{i,m}$, $T_{f,m}$, $T_{i,e}$ and $T_{f,e}$ according to the self-controlling measuring path that was proposed for gelled propellant manufacturing.

- Thermohygrograph: As a portable drum temperature and humidity recorder, a thermohygrograph should be applied for room temperature registration during the intermediate storage (in the case of gelled propellants that needs this auxiliary stage), what means the temperatures at the end of batch mixing $T_{f,m}$ and at the beginning of ram extrusion $T_{i,e}$. So the evolution of these process parameters is printed on graph paper thanks to the rotating drum. The drum rotation and data recording are carried out by a mechanical system without any electronic devices, and thus static electricity effects are avoided, ensuring safe conditions for process parameter inspection.
- Precision scale: The measuring of propellant dough density shall be made by a precision scale. Among other possible magnitudes related to the mass and volume of solid and semi-solid material samples, the material densities ρ_m and ρ_e can be obtained by using these instruments.
- Rotational rheometer: The rotational rheometers provide diverse techniques for the rheological testing of propellant doughs and other different semi-solid compounds, making possible to determine a great amount of rheological parameters that include the storage modulus G' and loss modulus G''. According to the proposed methodology for production monitoring, the oscillatory rheological moduli G'_m, G''_m, G'_e and G''_e described in a previous section can be measured by small-amplitude oscillatory shear (SAOS) tests, and thus the complex modulus for G^*_m and G^*_e that correspond to mixing and extrusion processes, respectively.
- Extrusion control devices: The main parameters of the extrusion stage can be measured by means of the control equipment of ram extrusion press. This equipment allows the registration of steady extrusion pressure p_e and ram velocity v_e during this manufacturing process.

5 Procedures and Registers for Propellant Production Controlling

In order to define an adequate quality control methodology for a gelled propellant production system, a series of procedures are needed to establish the routines to be followed for each measuring or evaluating action, as well as the template for registers to be saved with the data collected during the use of this methodology [8, 9]. For controlling and monitoring of propellant manufacturing, the following procedures should be created:

- General procedure: This document must describe the key factors of the quality control, including the aim, purpose and scope. The general procedure should contain a flow diagram about the control parameters to be considered, with the definition and coding for each one of these variables, a description of the self-controlling measuring path to be followed, and a list with the totality of quality control procedures to be applied.

- Procedure about measuring of mixing electrical power: This procedure must describe the installation of a power monitoring device on the power line of a batch mixer and its use to evaluate the instantaneous power consumption during the kneading process. The purpose of this document is to explain the mode of operation of these devices, in order to determine the mixing energy E_{m} from the power consumption and effective time.
- Procedure about measuring of material temperature: The instruments to be applied and the steps to be followed for measuring the temperature of propellant doughs during its processing must be explained in this procedure. This procedure shall include basic information about the pyrometer, how to use this instrument and the temperatures to be registered during the batch mixing and ram extrusion stages ($T_{\mathrm{i,m}}$, $T_{\mathrm{f,m}}$, $T_{\mathrm{i,e}}$ and $T_{\mathrm{f,e}}$).
- Procedure about measuring of material density: This document must describe the appropriate device for this purpose, and how it should be used to obtain with a high safety the density of propellant doughs ρ_{m} and ρ_{e}.
- Procedure about SAOS analysis: The basic instructions about the use of a rotational rheometer for measuring of rheological moduli of propellant doughs must be contained by this procedure. This document will provide guidelines about the preparation of material samples, execution of SAOS tests and determination of rheological moduli G' and G'' within the LVR, and thus the complex modulus G^*.
- Procedure about measuring of extrusion parameters: The routines to be followed for measuring the extrusion pressure p_{e} and ram velocity v_{e} are explained in this procedure. This document must contain the details about the specific control system implemented in the extrusion facility.
- Procedure about safety assessment: This procedure should describe the totality of actions for ensuring the safety requisites associated with the process quality control. The applicable issues about ATEX normative to be assumed for the production and measuring facilities shall be included in this procedure.

From the data measured by the proposed devices according to this control methodology, different registers containing the information about the gelled propellant manufacturing system should be saved. The data to be contained in these registers includes the production date, propellant name, batch number, raw material requirements (solubility, particle size, nitration temperature, etc.) and control parameters of this methodology ($T_{\mathrm{i,m}}$, $T_{\mathrm{f,m}}$, t_{m}, E_{m}, m, ρ_{m}, G'_m, G''_m, G^*_m, t_{s}, $T_{\mathrm{i,e}}$, $T_{\mathrm{f,e}}$, ρ_{e}, G'_e, G''_e, G^*_e, p_{e} and v_{e}). Figure 8 shows an example of the template that can be adopted to elaborate these registers.

These registers with the data related to each propellant formulation and manufacturing process can be saved separately, i.e. the registers for batch mixing or ram extrusion and the registers for simple-base, double-base and triple-base propellants.

Propellant name	Stage	RAW MATERIAL			BATCH MIXING								RAM EXTRUSION								
	Variable Code	RM.1	RM.2	RM.3	A.1.1	A.1.2	A.2.1	A.2.2	A.3	A.4.1	A.4.2	A.4.3	S.1	E.1.1	E.1.2	E.2.1	E.2.2	E.3	E.4.1	E.4.2	E.4.3
	Mean Value																				
	Std. Deviation																				
Date	Batch nº	S	p.s.	T_{nit}	$T_{i,m}$	$T_{f,m}$	t_m	E_m	ρ_m	G'_m	G''_m	G^*_m	t_s	$T_{i,e}$	$T_{f,e}$	v_e	p_e	ρ_e	G'_e	G''_e	G^*_e
mm/dd/yy	XXX	%	μm	°C	°C	°C	min	Wh	kg/m³	Pa	Pa	Pa	min	°C	°C	mm/s	bar	kg/m³	Pa	Pa	Pa

Fig. 8 Example of quality control register for a specific gelled propellant

The information contained in these registered should be accessible and editable, in order to make possible the checking, comparison and selection of the data assumed for specific production batches. For this purpose, a specific software should be employed to handle the registers of this database.

6 Example of Industrial Application of This Production Control Methodology

In this section, the implementation of this control methodology for batch mixing and ram extrusion during the gelled propellant production is illustrated by means of some examples.

During the batch mixing stage, the electric power consumption of the batch mixer must be registered employing a power monitoring device as previously mentioned. This equipment serves to obtain the curve of electric active power $P_m(t)$ as a function of mixing time. The curve depicted in Fig. 9 corresponds to a triple-base propellant.

The power peaks that can be seen in this figure are a consequence of the machine stops for adding the required chemical elements or checking the mechanical properties of the mixture, while a horizontal tendency is evidenced at the end of the process due to the chemical reactions that allow the gelation. This flat region is originated once the gelation of original compound is finished, and after that the kneading process must be hold during a certain interval in order to obtain the desired homogeneity in the propellant doughs.

The area below this curve provides the specific mixing energy E_m. When the required level of mixing energy is achieved, the mixture can be assumed to be ready for the forming process that must be carried out in the next stage of manufacturing system.

As previously described, the temperature increment in the propellant dough during the mixing process can be determined by measuring the temperature level in the core of resultant material, using a pyrometer with a penetration thermocouple.

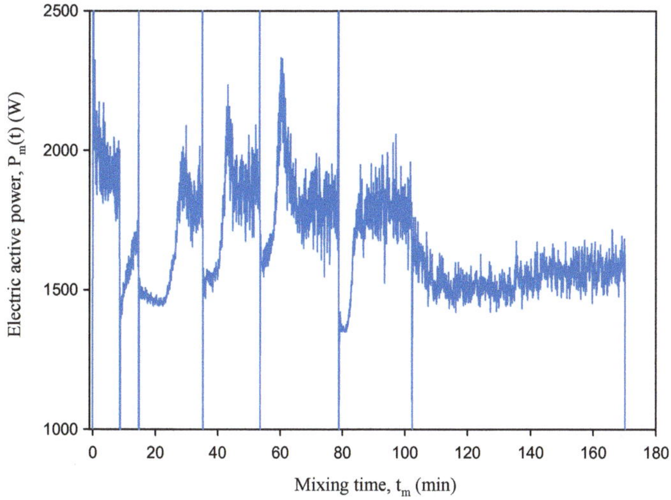

Fig. 9 Time-dependent electric power consumption during batch mixing of a triple-base propellant

The temperature measuring must be carried out at the beginning and at the end of each batch mixing.

The temperature of propellant doughs at the end of the mixing process will depend on distinct factors, e.g. the overall mixing time, the environmental temperature or the selected zone for temperature inspection inside the mixing machine. So the empirical correlation between the initial and final mixing temperatures could be difficult to obtained with sufficient accuracy, as can be seen for instance in Fig. 10.

From the experimental correlation between these parameters, the following expression can be deduced between the initial and final temperatures during the mixing process:

$$T_{\mathrm{f,m}} = a_1 + a_2 T_{\mathrm{i,m}} \tag{13}$$

where a_1 = 13.878 and a_2 = 0.762 for these propellant formulations.

The mechanical response of propellant dough can be evaluated by means of the storage modulus G'_m and loss modulus G''_m. These rheological parameters within the LVR can be determined using oscillatory amplitude sweep tests. About four or five samples should be measured by the rotational rheometer for each batch mixing, if reliable values of these rheological moduli must be obtained (Fig. 11).

Once four or five samples of the propellant doughs have been rheological characterized by means of oscillatory amplitude sweep tests (Fig. 11), for a given batch mixing and a certain value of E_{m}, the average values of each modulus within the LVR must be deduced. For example, the following values can be obtained in the case of a triple-base propellant:

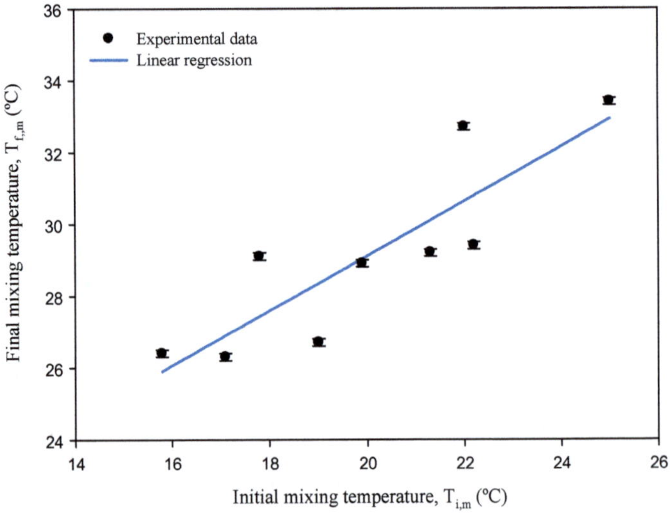

Fig. 10 Relationship between the initial and final temperatures during batch mixing of a double-base propellant

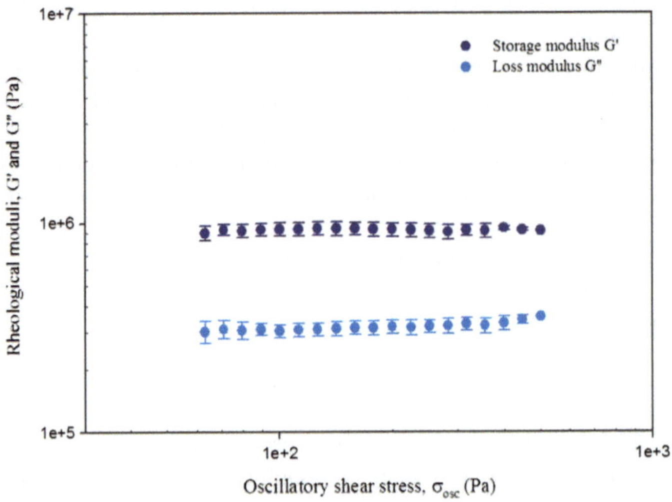

Fig. 11 Storage and loss moduli at the end of batch mixing for a double-base propellant

$$G'_{m} = 9.08 \times 10^5 \pm 0.91 \times 10^5 (\text{Pa}) \tag{14}$$

$$G''_{m} = 3.08 \times 10^5 \pm 0.23 \times 10^5 (\text{Pa}) \tag{15}$$

In addition, in order to simplify the process analysis and to deduce the possible empirical correlations, the viscoelastic state of the propellant doughs can be represented by a unique rheological parameter such as the complex modulus G^*. From the previous equations, the resultant value of complex modulus can be calculated as:

$$G^*_{\mathrm{m}} = 9.59 \times 10^5 \pm 0.93 \times 10^5 (\mathrm{Pa}) \quad (16)$$

This average value of the complex modulus G^*_m within the LVR can be employed to find out the possible relationship between this material parameter and the mixing energy (Fig. 12).

By means of empirical correlation, a quantitative relation between the viscoelastic state of mixed propellant doughs and the applied process parameters could be deduced. The higher the number of measured batch mixings, the higher will be the regression factor R^2 that represent the accuracy of these correlations. This analysis allows predicting the viscoelastic state of propellant doughs represented by G^*_m from the mixing conditions that are defined by mixing energy $E_{m,}$ and vice versa.

The following mathematical expression can be obtained by experimental correlation between both parameters:

$$G^*_{\mathrm{m}} = b_1 - b_2 E_{\mathrm{m}} \quad (17)$$

where b_1 = 2,504,562 and b_2 = 288 in the case of these propellants.

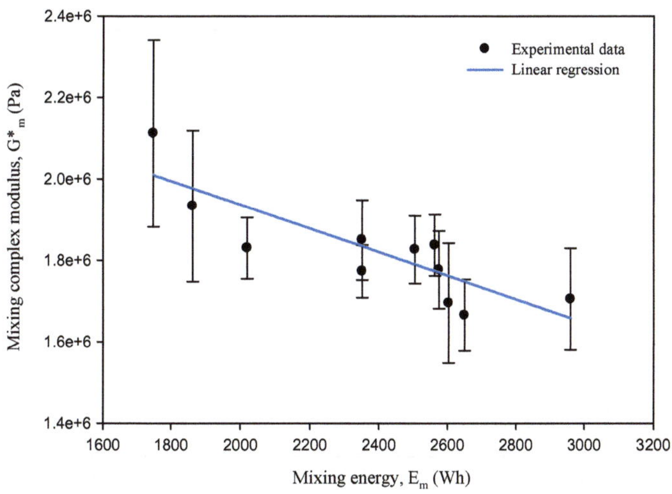

Fig. 12 Relationship between complex modulus and mixing energy during batch mixing of a double-base propellant

Once propellant doughs are obtained through the batch mixing process, depending on the type of propellant an intermediate storage stage will take place before the extrusion process. During this time (storage time), the viscoelastic state of such propellant doughs could be affected by a slight loss of solvents, which would provoke a certain increase in the complex modulus G^*.

The following step for an appropriate process control will correspond to the supervision of the selected pressure during the ram extrusion process. For this purpose, not a fixed value but a time-dependent function must be established for extrusion pressure.

If the instantaneous values of extrusion pressure are plotted as function of the processing time (Fig. 13), a first zone with an increasing transient pressure and a second zone with an average steady pressure can be clearly identified. For empirical correlation with other process parameters, this average value of steady extrusion pressure p_e must be considered.

In this figure, the dotted line represents the pressure level during the steady period of the extrusion process p_e, after the initial established pressure ramp is applied by the control system of the extrusion press to reach the desired processing conditions, which cannot be previously predicted if the value of G_e^* is unknown. If G_e^* was not used for to the control process, the steady pressure level should be adjusted by the operators based on their experience.

However, in cold extrusion processes, the steady extrusion pressure p_e can be identified as a function of the viscoelastic state of propellant doughs, i.e. rheological moduli G_e' and G_e'' and G_e'', and the influence of temperature can be neglected in comparison to this other parameters. If the extrusion pressure p_e is represented

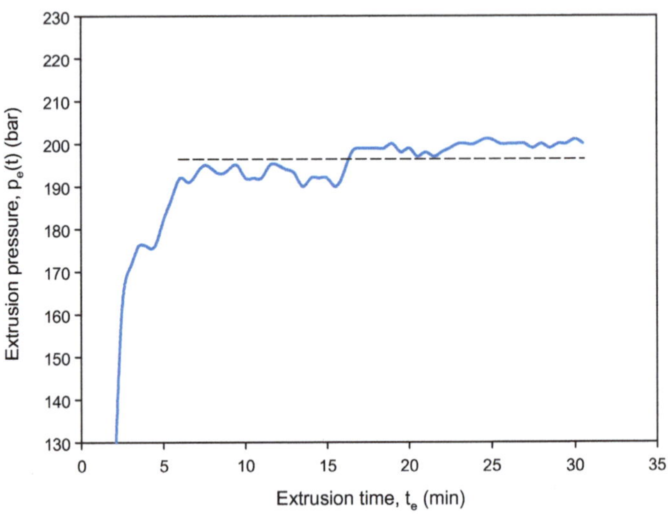

Fig. 13 Time-dependent extrusion pressure during ram extrusion of a double-base propellant

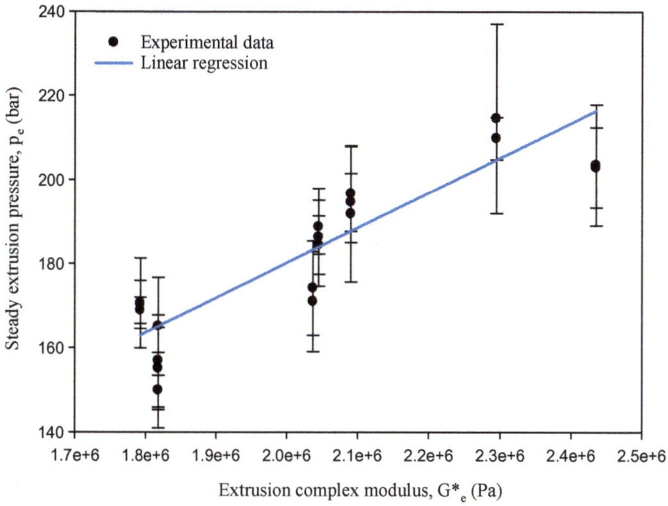

Fig. 14 Relationship between extrusion pressure and complex modulus during ram extrusion of a double-base propellant

against the complex modulus of extrusion process G_e^*, the relationship between both parameters could be identified (Fig. 14).

The relationship between G^* and p_e is crucial for this quality control methodology, since it makes possible to predict the pressure ramps and final steady extrusion pressure that would be needed according to the complex modulus of compound to be subject to extrusion process G_e^*, which would be previously measured by oscillatory amplitude sweeps tests.

The steady pressure can be related to the complex modulus by the following expression, according to the empirical correlation between both parameters:

$$P_\mathrm{e} = c_1 + c_2 G_\mathrm{e}^* \tag{18}$$

where c_1 = 21.47 and c_2 = 7.9 × 10^5 for these gelled propellants.

The variability associated with the ram extrusion process (expressed in terms of the dispersion of extrusion pressure p_e) can be attributed to the dispersion of the viscoelastic behaviour shown by the kneaded propellant doughs, which depends on the degree of gelation reached during the batch mixing stage [5]. So the appropriate selection of the sequences for adding each compound, the period of gelation and the overall duration of mixing process can be carried out to reduce the dispersion in the viscoelastic state of propellant doughs.

The steady pressure level for extrusion process p_e is strongly related to the imposed ram velocity v_e. For this reason, the pressure level to be assumed must the adjusted continuously to guarantee the desired ram velocity during the production process [32, 33].

When dealing with cold extrusion, no temperature level is imposed during the process, then isothermal conditions are assumed. Nevertheless, the viscous heating is always generated due to wall friction effects along the extrusion press surfaces [15, 30, 31]. In this sense, the initial and final temperatures of the material should be measured not only to ensure that the isothermal conditions are an adequate assumption, but also to determine the temperature rising occurred during the ram extrusion of each type of propellant [16, 17] (Fig. 15).

A different temperature increment can be originated for each propellant formulation, which would be useful to decide the assumptions to be adopted about thermal effects, to develop new formulations and to introduce certain modifications on the propellant composition.

The relationship between the initial and final extrusion temperatures can be described by means of the following expression:

$$T_{\mathrm{f,e}} = d_1 + d_2 T_{\mathrm{i,e}} \tag{19}$$

where d_1 = 4.123 and d_2 = 0.936 for these propellants' formulations.

Finally, an additional parameter could be also included in this quality control, i.e. the final diameter as representative dimension of the extruded product cross section. This additional parameter can be measured inside the quality control procedure. This parameter will depend on several factors, e.g. extrusion die design and die swell, the existence of drying or vacuum devices, the traction exerted by the pulling system.

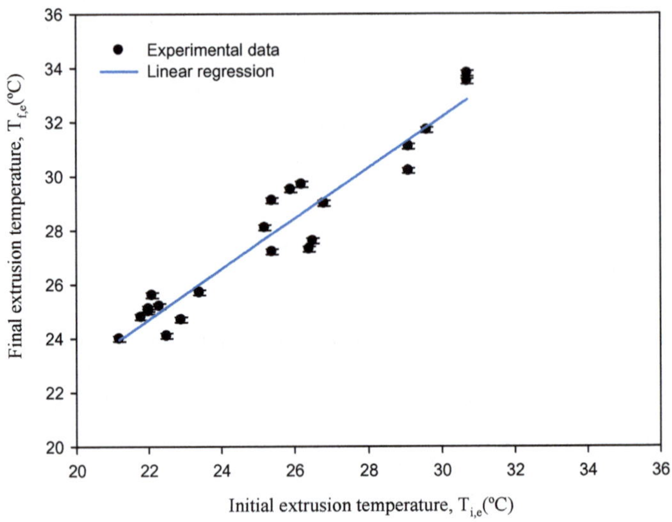

Fig. 15 Relationship between initial and final temperatures during ram extrusion of a double-base propellant

7 Conclusions

This chapter is focused on the fundamentals of quality and productivity improvement in the manufacturing of gelled propellants. For this purpose, the methodology to be applied in the processes used to produce these energetic materials is described, including the measuring devices that should be implemented to acquire the experimental data, the testing techniques for characterization of rheological properties and the influence of process and material parameters on the product quality. In this chapter, the procedures for quality control controlling batch mixing and ram extrusion are explained, as some of the main processes for manufacturing of gelled propellants. These principles could also serve as a reference for other production techniques like screw extrusion.

References

1. Urbanski T (1986) Chemistry and technology of explosives. Pergamon Press, New York
2. Meyer R (2007) Explosives. Wiley-VCH, Weinheim
3. Teipel U (2005) Energetic materials: particle processing and characterization. Wiley-VCH, Weinheim
4. Strütt H (1996) Handbook of mixing technology. Ekato, Schopfheim, Germany
5. Kalyon DM, Dalwadi D, Erol M, Birinci E, Tsenoglu C (2006) Rheological behavior of concentrated suspensions as affected by the dynamics of the mixing process. Rheol Acta 45(5):641–658
6. Yazici R, Kalyon DM, Fair D (1998) Microstructure and mixing distribution analysis in M30 triple-base propellants. Stevens Institute of Technology, Highly Filled Materials Institute, Hoboken
7. Martinez-Pastor J, Franco P, Ramirez FJ, Lopez-Garcia P (2016) Influence of rheological behaviour on extrusion parameters during non-continuous extrusion of multi-base propellants. IntJ Mater Form. doi:10.1007/s12289-016-1332-5
8. Evans JR, Lindsay WM (2002) The management and control of quality, 5th edn. South-Western, Cincinnati, Ohio
9. Groover MP (2007) Fundamentals of modern manufacturing: materials processes, and systems. Wiley, New York
10. Baker FS, Healey MJ, Privett G (1988) The rheological properties of plasticized nitrocellulose as a function of nitrocellulose precursor. Propellants Explos Pyrotech 13(4):99–102
11. Baker FS, Carter RE (1982) Processing characteristics of gun propellants. Propellants Explos Pyrotech 7(6):139–147
12. Birinci E, Gevgilili H, Kalyon DM, Greenberg B, Fair DF, Perich A (2006) Rheological characterization of nitrocellulose gels. J Energ Mater 24(3):247–269
13. Giles HF, Wagner JR, Mount EM (2005) Extrusion: the definitive guide processing guide and handbook. William Andrew Publishing, New York
14. Michaeli W (1992) Extrusion dies for plastics and rubber. Hanser Publishing, Munich
15. Carter RE, Warren RC (1987) Extrusion stresses, die swell, and viscous heating effects in double-base propellants. J Rheol 31(2):151–173
16. Lawal A, Kalyon DM (1997) Viscous heating in nonisothermal die flows of viscoplastic fluids with wall slip. Chem Eng Sci 52(8):1323–1337
17. Lawal A, Kalyon DM (1997) Nonisothermal extrusion flow of viscoplastic fluids with wall slip. Int J Heat Mass Transf 40(16):3883–3897

18. Lieb RJ (1987) Nitroguanidine morphology in extruded gun propellant. Technical report BRL-TR-2812, Army Ballistic Research Lab Aberdeen Providing Ground MD
19. Yilmazer U, Kalyon DM (1989) Slip effects in capillary and parallel disk torsional flows of highly filled suspensions. J Rheol 33(8):1197–1212
20. Kalyon DM, Yaras P, Aral B, Yilmazer U (1993) Rheological behavior of a concentrated suspension: a solid rocket fuel simulant. J Rheol 37(1):35–53
21. Kalyon DM (2005) Apparent slip and viscoplasticity of concentrated suspensions. J Rheol 49(3):621–640
22. Rahimi S, Peretz A, Natan B (2007) On shear rheology of gel propellants. Propellants Explos Pyrotech 32(2):165–174
23. Martinez-Pastor J, Franco P, Ramirez-Fernandez FJ (2014) Rheological characterization of energetic materials by rotational testing techniques. In: ASME 2014, 12th biennial conference on engineering systems design and analysis, 25–27 June 2014, Copenhagen, Denmark
24. Barnes HA (1993) An introduction to rheology. Elsevier Science Publishers B.V, Amsterdam
25. Barnes HA (2000) A handbook of elementary rheology. University of Wales, Aberystwyth, UK
26. Macosko CW (1994) Rheology: principles, measurements and applications. Wiley-VCH, New York
27. Tanner RI (2002) Engineering rheology. Oxford Engineering Science Series, New York
28. Mitsoulis E, Abdali SS, Markatos NC (1993) Flow simulation of Herschel-Bulkley fluids through extrusion dies. Can J Chem Eng 71(1):147–160
29. Mitsoulis E (2007) Annular extrudate swell of pseudoplastic and viscoplastic fluids. J Nonnewton Fluid Mech 141(2):138–147
30. Lawal A, Kalyon DM, Yilmazer U (1993) Extrusion and lubrication flows of viscoplastic fluids with wall slip. Chem Eng Commun 122(1):127–150
31. Rahimi S, Durban D, Khosid S (2010) Wall friction effects and viscosity reduction of gel propellants in conical extrusion. J Nonnewton Fluid Mech 165(13):782–792
32. Warren RC, Starks AT (1988) Multiple flow regimes in the extrusion of nitrocellulose based propellant doughs. Technical report WSRL-TR-43/88
33. Denn MM (2001) Extrusion instabilities and wall slip. Annu Rev Fluid Mech 33(1):265–287

Random Excitation Technique for Measurement of Acoustic Properties

Gaurav Sharma and Arpan Gupta

1 Introduction

The acoustic properties of a test specimen can be calculated using a randomly excited signal through a duct and measuring the properties of signal reflected from the test specimen. The various stages of the process include defining the objective, understanding relevant theory, identification of measurable parameters, formulation of theory and instrumentation for measurement, accuracy enhancement, controlling parameters and, most importantly, limitations of the used procedure. A random excitation technique for measurement of acoustic properties, with intermediate processes undergone, is presented in the following text.

To perform a study or analysis in a particular field, familiarity of the literature is required. For that, relevant literature review or Acoustic Wave Theory is presented for the reader which may be referred to, from time to time depending on the requirement in further sections. Also, the necessary mathematics is presented. The accuracy, cost, time, etc. of a particular analysis are highly influenced by the type of experimental set-up chosen. Various set-ups that can be used for the current analysis are described and compared, and a particular set-up is selected. The theory for the selected set-up is formulated according to the requirements.

Even after having an accurate set-up, proper selection of the measurement devices is required, or otherwise, the results obtained may be highly erroneous. So, various instrument parameters (specifications) are taken care of, while selecting the measurement devices. Some of them are discussed in detail. The signals can be pre-amplified to enhance signal characteristics. The signals acquired by most of the

G. Sharma · A. Gupta (✉)
Indian Institute of Technology, Mandi 175005, Himachal Pradesh, India
e-mail: agupta@iitmandi.ac.in

G. Sharma
e-mail: ggsharma93@gmail.com

M. Ram and J.P. Davim (eds.), *Modeling and Simulation in Industrial Engineering*,
Management and Industrial Engineering, DOI 10.1007/978-3-319-60432-9_2

measurement devices are in the analog form. But, for analysis by computing devices, the signal needs to be converted into digital form. The wrong selection of A-to-D converters leads to introduction of error at this least expected stage. Depending on the type of data, the best suited A-to-D converter changes. The finally acquired data can be processed. But there are some common sources of error which need to be taken care of. Some basic precautions that can be helpful in eliminating these errors are mentioned.

2 Acoustic Waves

Acoustic waves or sound waves are longitudinal waves which travel by means of repetitive compression and expansion [1]. As the sound waves travel at sufficiently high speeds, the process is assumed to be adiabatic in nature. These waves exhibit phenomena like interference, diffraction and reflection which need to be quantified, so as to predict their behaviour on interaction with various materials and use them to modify the acoustic properties in various environments like reduction of reverberation time in large enclosures for audibility, design of audio instruments, muffler designs, sound proofing and many more applications.

A comparatively easier study of acoustic waves is possible, if wave front of the acoustic waves is planar. These types of wave are called plane waves.

2.1 *Plane Waves*

Plane waves may be referred to as waves with a planar wave front, i.e. the magnitude and phase of pressure perturbation (p) and particle velocity (u) are same on the plane perpendicular to the direction of wave propagation. This type of wave is generated in a tubular duct when the wave travels along its longitudinal axis. The basic equation (1D) for linearization can be achieved by taking elementary mass (Fig. 1), through which the acoustic wave is travelling,where

Elementary volume be $\mathrm{d}m$
Ambient pressure p_0 and density ρ_0
Acoustic perturbations on pressure p and on density ρ
Control volume $\mathrm{d}V = \mathrm{d}x\mathrm{d}y\mathrm{d}z$
For acoustic waves $p/p_0 1, \rho/\rho_0 1$

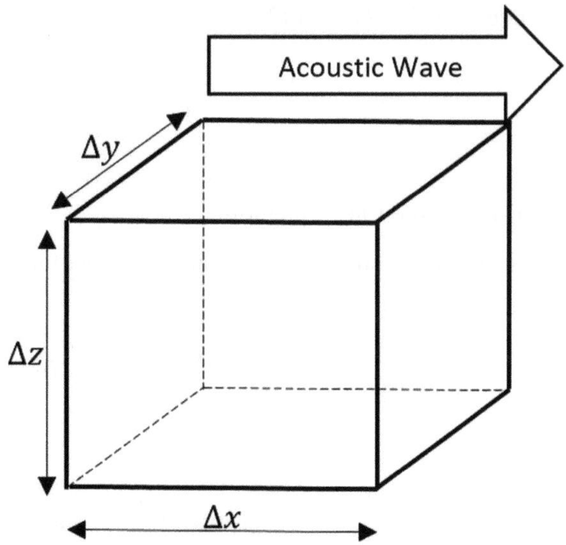

Fig. 1 Control volume showing acoustic wave flow in x-direction

- **Conservation of Mass**

Mass inflow

$$dm_{in} = \rho_0 u \, dydz$$

Mass outflow

$$dm_{out} = \left(\rho_0 u + \frac{\partial(\rho_0 u)}{\partial x} dx\right) dydz$$

Net mass increase in the control volume

$$dm_{in} - dm_{out} = \rho_0 u \, dydz - \left(\rho_0 u + \frac{\partial(\rho_0 u)}{\partial x} dx\right) dydz$$

or

$$dm_{in} - dm_{out} = -\frac{\partial(\rho_0 u)}{\partial x} dxdydz \tag{1}$$

Net change in density of control volume can be written as

$$\frac{\partial \rho}{\partial t} dxdydz = dm_{in} - dm_{out} \tag{2}$$

Equating Eqs. (1) and (2)

$$\frac{\partial \rho}{\partial t} + \rho_0 \frac{\partial u}{\partial x} = 0 \tag{3}$$

- **Conservation of Momentum**

Net change in momentum is equal to the net force acting on the control volume, i.e.

$$\frac{\partial(\rho_0 u \mathrm{d}x\mathrm{d}y\mathrm{d}z)}{\partial t} = p_0 \mathrm{d}y\mathrm{d}z - (p_0 + \mathrm{d}p)\mathrm{d}y\mathrm{d}z$$

or

$$\rho_0 \mathrm{d}x\mathrm{d}y\mathrm{d}z \frac{\partial u}{\partial t} = -\mathrm{d}p\mathrm{d}y\mathrm{d}z$$

or

$$\rho_0 \frac{\partial u}{\partial t} = -\frac{\mathrm{d}p}{\mathrm{d}x}$$

or

$$\rho_0 \frac{\partial u}{\partial t} + \frac{\mathrm{d}p}{\mathrm{d}x} = 0$$

As distance x and time t are mutually independent variables,

$$\rho_0 \frac{\partial u}{\partial t} + \frac{\partial p}{\partial x} = 0 \tag{4}$$

- **Conservation of Energy (Adiabatic Equation)**

For an adiabatic process,

$$\frac{p + p_0}{(\rho + \rho_0)^{\gamma}} = \text{constant (say } C\text{)}$$

Taking log both sides,

$$\log(p+p_0) - \gamma(\log(\rho+\rho_0)) = \log(C_1)$$

Differentiating with respect to ρ,

$$\frac{1}{p+p_0}\frac{\partial p}{\partial \rho} - \gamma\frac{1}{\rho+\rho_0} = 0$$

or

$$\frac{\partial p}{\partial \rho} = \gamma\frac{p+p_0}{\rho+\rho_0} \cong \gamma\frac{p_0}{\rho_0} = c_0^2\,(\text{say}) \tag{5}$$

Equation (5) implies

$$\mathrm{d}\rho = \frac{\mathrm{d}p}{c_0^2};\quad \frac{\partial \rho}{\partial t} = \frac{1}{c_0^2}\frac{\partial p}{\partial t};\quad \frac{\partial \rho}{\partial x} = \frac{1}{c_0^2}\frac{\partial p}{\partial x} \tag{6}$$

Substituting Eq. (6) in (3),

$$\rho_0\frac{\partial u}{\partial x} + \frac{1}{c_0^2}\frac{\partial p}{\partial t} = 0$$

Differentiating with respect to t,

$$\rho_0\frac{\partial^2 u}{\partial x \partial t} + \frac{1}{c_0^2}\frac{\partial^2 p}{\partial t^2} = 0 \tag{7}$$

Differentiating Eq. (4) with respect to x,

$$\rho_0\frac{\partial^2 u}{\partial x \partial t} + \frac{1}{c_0^2}\frac{\partial^2 p}{\partial x^2} = 0 \tag{8}$$

Subtracting Eq. (7) from (8),

$$\frac{\partial^2 p}{\partial t^2} - c_0^2\frac{\partial^2 p}{\partial x^2} \tag{9}$$

This type of equation with homogeneous form and constant coefficients (c_0) has a general form of solution

$$p(x,t) = C_1 f\left(t - \frac{x}{c_0}\right) + C_2 f\left(t + \frac{x}{c_0}\right) \tag{10}$$

For proceeding to the solution of the above differential equation, let us assume time dependence to be of the exponential form $e^{\imath\omega t}$. Then, the solution becomes

$$p(x,t) = C_1 e^{\imath\omega\left(t-\frac{x}{c_0}\right)} + C_2 e^{\imath\omega\left(t+\frac{x}{c_0}\right)} \tag{11}$$

The first part repeats itself at $x = c_0 t$, i.e. forward travelling wave with velocity c_0, and second part of the same as backward travelling wave with same velocity. Therefore, c_0 is the wave velocity and C_1 and C_2 are the amplitudes of the forward and backward travelling waves.

Equation (11) can be rearranged as

$$p(x,t) = [A e^{-\imath kx} + B e^{\imath kx}] e^{\imath\omega t} \tag{12}$$

where $k = \omega/c_0$, A is the amplitude of forward travelling wave and B is the amplitude of backward travelling wave. From Eq. (4),

$$-\rho_0 \frac{\partial u}{\partial t} = \frac{\partial p}{\partial x} = -\rho \frac{\partial^2 \phi}{\partial x \partial t} \text{ (say)}$$

or

$$u = \frac{\partial \phi}{\partial x}; \quad p = -\rho_0 \frac{\partial \phi}{\partial t} \tag{13}$$

where ϕ is referred as velocity potential function. Therefore, the dependent variable in the wave equation may be ϕ or u instead of p. So that same solution can be generated for u, i.e.

$$u(x,t) = [C_3 e^{-\imath kx} + C_4 e^{\imath kx}] e^{\imath\omega t} \tag{14}$$

Substituting Eqs. (12) and (14) in (4),

$$C_3 = \frac{A}{\rho_0 c_0}, \quad C_4 = -\frac{B}{\rho_0 c_0}$$

And therefore,

$$u(x,t) = \frac{1}{Z_0} \left[A e^{-\imath kx} - B e^{\imath kx}\right] e^{\imath\omega t} \tag{15}$$

where $Z_0 = \rho_0 c_0$, the characteristic impedance (later referred as Y) of the material and S is the cross section of tube.

$$Y = \begin{cases} \rho_0 c_0 & \text{for particle velocity } u, \\ \frac{(\rho_0 c_0)}{S} & \text{for volume velocity } \upsilon_\upsilon, \\ \frac{c_0}{S} & \text{for mass velocity } \upsilon. \end{cases} \tag{16}$$

For 3D waves in an inviscid medium,

$$\frac{\partial^2 p}{\partial t^2} - c_0^2 \nabla^2 p = 0 \tag{17}$$

where ∇^2 is the 3D Laplacian operator.

2.1.1 Waves in Absorptive Stationary Medium

Not going much further into detail, the particle motion with heat conduction from walls of circular tube induces radial motion of particles. Due to the presence of viscosity, the radial and axial motion of the particles can be assumed to be independent.

After a lot of mathematics [1], the final results for wave number for a progressive wave in a duct can be written as

$$k = k_0 + \alpha \tag{18}$$

where $k_0 = \omega / c_0$ is the wave number in free medium and α is the attenuation constant.

The standing wave solution for ducts can be shown as

$$p(x,t) = \left[A\mathrm{e}^{-\alpha x - \imath kx} + B\mathrm{e}^{\alpha x + \imath kx} \right] \mathrm{e}^{\imath \omega t} \tag{19}$$

and the acoustic mass velocity,

$$\upsilon(x,t) = \frac{1}{Y} \left[A\mathrm{e}^{-\alpha x - \imath kx} - B\mathrm{e}^{\alpha x + \imath kx} \right] \mathrm{e}^{\imath \omega t} \tag{20}$$

where Y is the characteristic impedance for the forward travelling wave, where Y_0 is the characteristic impedance for inviscid medium given in Eq. (16),

$$Y = Y_0 \left(1 - \frac{\alpha}{k_0} + \imath \frac{\alpha}{k_0} \right) \tag{21}$$

3 Acoustic Waves

For the measurement of a property (acoustic impedance) in an experimental set-up, it is very essential to correlate it to the properties which can be measured easily, accurately and cost-effectively. In our case, the sound pressure level can be easily measured using a microphone and, when measured over a period of time, is sufficient to indicate the properties of an acoustic wave at a point. Various experimental set-ups exist to correlate the sound pressure level of a region to acoustic properties like impedance and reflection coefficient of the medium.

3.1 Set-up Selection

The experiment for the measurement of acoustic impedance of a sample surface or membrane can be done using various experimental set-ups, each having individual advantages and disadvantages.

3.1.1 Sliding Microphone Probe Experiment (SME)

A sketch of the B&K Standing Wave Apparatus is shown in Fig. 2 [2]. A loudspeaker is mounted on one end of the acoustic tube and is used to produce acoustic wave. This wave gets reflected from the surface of the test sample, mounted on the other end of the impedance tube. If rigid wall is present on the end surface of the sample, the magnitude of reflection coefficient is set to unity. The sliding microphone probe is moved along the impedance tube to measure minimum and maximum pressure corresponding to maxima and minima of the standing wave generated inside the tube due to incident and reflected wave components. The ratio of maximum to minimum pressure is known as SWR (standing wave ratio). The positions and magnitudes of these maxima and minima can be used to measure acoustic impedance and absorption coefficient of the test surface.

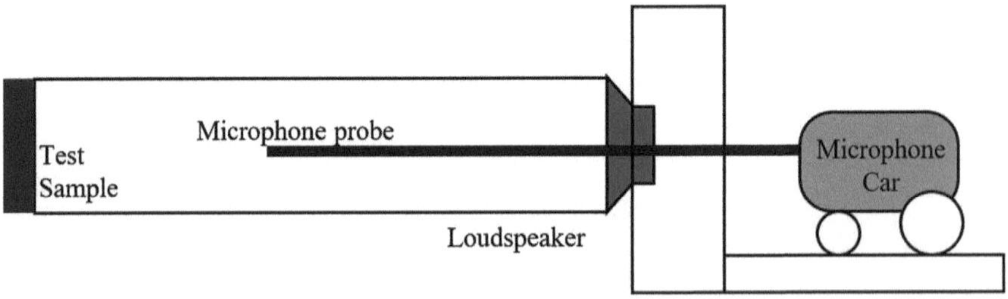

Fig. 2 Sliding microphone probe apparatus

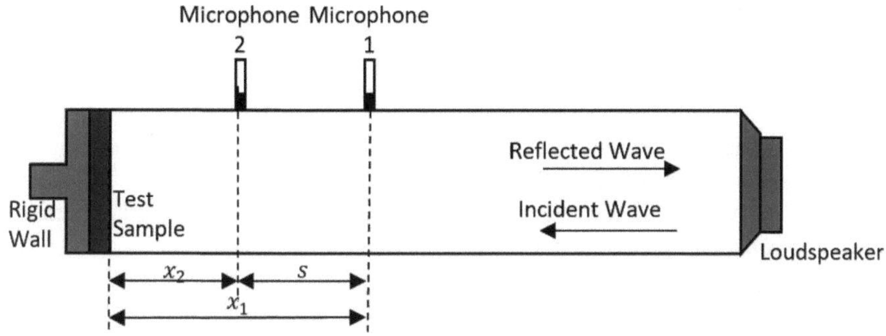

Fig. 3 Fixed microphone impedance tube apparatus

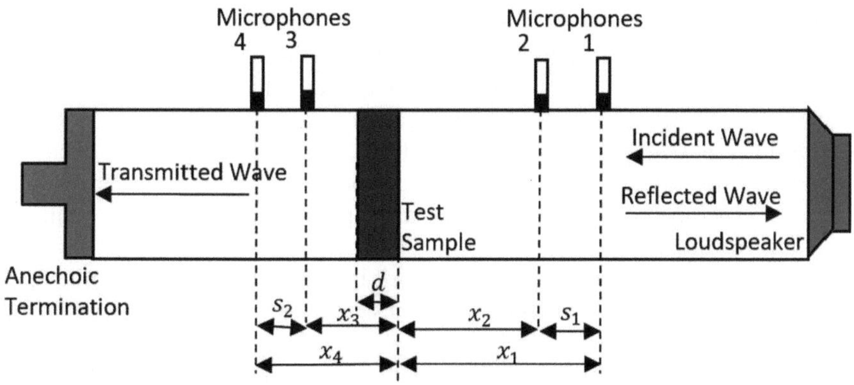

Fig. 4 Transfer matrix method apparatus

3.1.2 Fixed Two Microphone Impedance Tube Experiment (FME)

This method makes use of two microphones located at fixed positions [1]. The excitation can be random or discrete, depending on the requirement. A speaker is used to produce the acoustic wave inside the impedance tube. The signal picked by each microphone at the two fixed locations is used to calculate acoustic properties of the test sample. The auto- and cross-spectral densities of the individual signals at the two microphone locations are used to back-calculate the auto- and cross-spectral densities of the incident and reflected wave. These values are further processed to evaluate the required acoustic properties. A sample set-up diagram used for this experiment is shown in Fig. 3 [3].

3.1.3 Transfer Matrix Method (TMM)

This method is a slight modification of the fixed two microphone impedance tube experiment. An additional set of microphones is mounted on the other side of test sample [4]. The hard boundary or rigid wall is replaced with anechoic termination. A schematic is shown in Fig. 4. The readings from the four microphones are used to calculate the values of transfer matrix elements which can be further used to evaluate acoustic impedance and other properties of the test sample.

3.1.4 Comparison

The above methods have various advantages and disadvantages [1]:

- SME needs very simple instrumentation as only SPLs (sound pressure levels) need to be measured.
- SME is more accurate than other set-ups as the usual data acquisition and retrieval errors are absent.
- Any signal or combination of signals can be used in FME and TMM, but only sinusoidal signal is to be provided in SME with sufficient strength, to ensure SPL at pressure minima sufficiently higher than ambient noise.
- For different frequencies, in TMM and FME, the same set-up can be excited at different frequencies and data recorded can be used to measure impedance in a single experiment. But different set of readings are required for each frequency in case of SME.
- The acoustic field is likely to be disturbed by the moving probe tube in SME, while wall-mounted microphones and transient testing methods do not interfere with the acoustic field.
- In FME and TMM, tube length does not limit the measurement. But SME needs tube length at least one wavelength of the lowest frequency of interest.
- A more elaborate set-up is required for TMM than SME or FME, i.e. two extra microphones.
- The TMM approach is numerically more expensive than SME or FME.

For the current analysis, FME is used.

3.2 Theory

The sound waves are longitudinal waves which can be mathematically represented as [1]

$$p^{(r)} = p^{+}\mathrm{e}^{\imath\omega t}\left[\mathrm{e}^{\imath kx^{(r)}}\mathrm{e}^{\alpha x^{(r)}} + |R|\mathrm{e}^{-\imath\theta}\mathrm{e}^{-\imath kx^{(r)}}\mathrm{e}^{-\alpha x^{(r)}}\right] \tag{22}$$

and

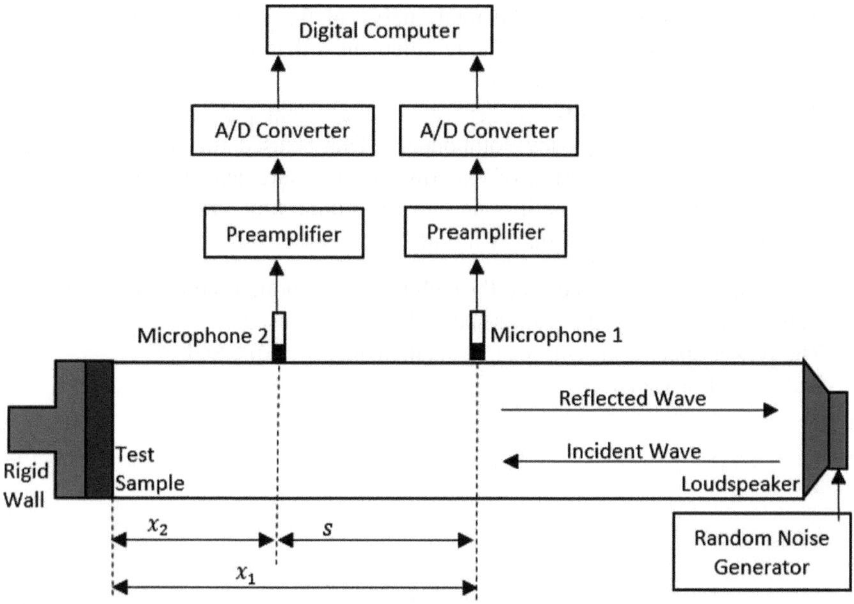

Fig. 5 Experimental set-up for fixed two microphone method

$$|p^{(r)} = |p^{+}|\left[\mathrm{e}^{2\alpha x^{(r)}} + |R|^2 \mathrm{e}^{-2\alpha x^{(r)}} + 2|R|\cos(\theta - 2kx^{(r)})\right] \quad (23)$$

where

p^{+}	is the incident wave sound pressure level.
$x^{(r)}$	is the distance measured from reflective surface ($x = 0$) in the direction opposite to that of the incident wave at the rth position (Fig. 5).
$\|R\|$ and θ	are amplitude and phase of reflection coefficient of the test sample.
α	is the tube attenuation and k is the wave number for stationary medium in the tube.

3.2.1 Spectral Density

The formulation of the experiment will later involve the calculation of spectral densities of the acquired signal. So, introductory text is provided for reference and better understanding of the current problem being analysed.

It is used to express the signal as distribution with the frequency variation [5]. The spectral density can be expressed as the distribution of energy or power of the signal with respect to frequency variation. The same can be classified on the basis of the quantity used for expressing distribution:

- **Energy Spectral Density** The total energy of the signal is decomposed into frequency components. The basic requirement is that the energy of the signal is finite, i.e. the signal is concentrated over a finite time interval.
- **Power Spectral Density** The energy of the signal per unit time interval is decomposed into frequency components. This is used for signals distributed over the whole time period of experimental measurement, i.e. if the signal is extended on the time axis, it is present for infinite time interval, and therefore, the total energy of the signal is infinite.

As the acoustic signal acquired through the microphone is continuous during the measured time interval, the power spectral density is used.

The power spectral density can be mathematically expressed as

Auto-Spectral Density

$$S_{AA}(f) = \frac{A(f)A^*(f)}{N^2} \tag{24}$$

Cross-Spectral Density

$$S_{AB}(f) = \frac{A(f)B^*(f)}{N^2} \tag{25}$$

where

N	is the number of points acquired in the time domain signal.
$A(f)$ and $B(f)$	are the fast Fourier transform (FFT) of time series signal $a(t)$ and $b(t)$, respectively, sampled for time period T. These are defined as

$$A(f) = \frac{1}{T}\int_0^T a(t)e^{\iota 2\pi ft} \tag{26}$$

and

$$B(f) = \frac{1}{T}\int_0^T b(t)e^{\iota 2\pi ft} \tag{27}$$

3.2.2 Evaluation of Attenuation Constant

For rigid case termination,

$$|R| = 1 \quad \text{and} \quad \theta = 0$$

The wave number can be calculated as

$$k = k_0 + \alpha = \frac{\omega}{c_0} + \alpha$$

where k_0 is the free medium wave number, $c_0 (= \sqrt{\gamma RT})$ is free medium wave velocity (velocity of sound) and ω is the dominant angular frequency of incident wave.

Substituting the above values in Eq. (23),

$$\begin{aligned} \left|p^{(r)}\right| = |p^+| \Big[& e^{2\alpha x^{(r)}} + e^{-2\alpha x^{(r)}} + 2\cos(2k_0 x^{(r)})\cos(2\alpha x^{(r)}) \\ & -2\sin(2k_0 x^{(r)})\sin(2\alpha x^{(r)}) \Big] \end{aligned} \tag{28}$$

where $r = 1, 2, 3\ldots$ As α is of the order 10^{-2} for a tube up to 1 m, one may simplify the above expression and retain only up to second-degree terms of $2\alpha x^{(r)}$, in the series expansion.

Thus,

$$\begin{aligned} \left|p^{(r)}\right|^2 = 2|p^+| \Big[& 1 + 2\alpha^2 (x^{(r)})^2 \Big\{ 1 - \cos(2k_0 x^{(r)}) \Big\} + \cos(2k_0 x^{(r)}) \\ & - 2\alpha x^{(r)} \sin(2k_0 x^{(r)}) \Big] \end{aligned} \tag{29}$$

Defining

$$\delta_{i,j} = \mathrm{SPL}_i - \mathrm{SPL}_j = 10\log_{10}\left|\frac{p^{(i)}}{p^{(j)}}\right|^2 = 10\log_{10}\beta_{i,j} \tag{30}$$

where $i, j = 1, 2, 3\ldots$

Here, $\beta_{i,j}$ can be evaluated from actual experimental values of level difference $\delta_{i,j}$.

From Eq. (29) and (30),

$$\beta_{i,j} = \frac{2\alpha^2 (x^{(i)})^2 [1 - \cos(2k_0 x^{(i)})] - 2\alpha x^{(i)} \sin(2k_0 x^{(i)}) + [1 + \cos(2k_0 x^{(i)})]}{2\alpha^2 (x^{(j)})^2 [1 - \cos(2k_0 x^{(j)})] - 2\alpha x^{(j)} \sin(2k_0 x^{(j)}) + [1 + \cos(2k_0 x^{(j)})]} \tag{31}$$

This equation, on rearrangement, reduces to a quadratic equation as

$$\alpha^2 - b_{i,j}\alpha + \frac{1}{2}c_{i,j} = 0 \ \text{ for } \ i, j = 1, 2, 3\ldots \tag{32}$$

where

$$b_{i,j} = \frac{\beta_{i,j} x^{(j)} \sin(2k_0 x^{(j)}) - x^{(i)} \sin(2k_0 x^{(i)})}{\beta_{i,j}(x^{(j)})^2[1 - \cos(2k_0 x^{(j)})] - (x^{(i)})^2[1 - \cos(2k_0 x^{(i)})]}$$

and

$$c_{i,j} = \frac{\beta_{i,j}[1 + \cos(2k_0 x^{(j)})] - [1 + \cos(2k_0 x^{(i)})]}{\beta_{i,j}(x^{(j)})^2[1 - \cos(2k_0 x^{(j)})] - (x^{(i)})^2[1 - \cos(2k_0 x^{(i)})]}$$

For the above set-up, $i = 1$ and $j = 2$.

For every pair of positions (i,j), this quadratic yields two possible roots for α. The positive root is the required one. If both the roots are positive, then solution of a number of such quadratic equations formed from SPL observations at a number of points arranged in pairs (i,j) would be necessary to pick out the correct value of α as the one that repeats itself consistently.

3.2.3 Normalized Impedance and Reflection Coefficient

After determination of α and k, the sound pressure can be denoted as

$$p(x,t) = A(f)e^{\iota(\omega t - kx)} + B(f)e^{\iota(\omega t + kx)} \tag{33}$$

and

$$\upsilon(x,t) = \frac{1}{Y}[A(f)e^{\iota(\omega t - kx)} - B(f)e^{\iota(\omega t + kx)}] \tag{34}$$

Therefore,

$$p(0,t) = \{A(f) + B(f)\}e^{\iota\omega t} \tag{35}$$

and

$$\upsilon(0,t) = \frac{1}{Y}\{A(f) - B(f)\}e^{\iota\omega t} \tag{36}$$

where A and B are functions of frequency (f) and attenuation is ignored.

If p and υ are random functions of time, then the required impedance $Z(\omega)$ or $Z(f)$ at $x = 0$, normalized with respect to the characteristic impedance of tube Y, is

$$\frac{Z(f)}{Y} = Z_n(f) = \frac{S_{p\upsilon}(f)}{YS_{\upsilon\upsilon}(f)} \tag{37}$$

where

$S_{pv}(f)$ cross-spectral density between p and υ at $x = 0$.
$S_{\upsilon\upsilon}(f)$ auto-spectral density of υ at $x = 0$.

And these, in turn, are related to FFT (fast Fourier transform) of p and υ at $x = 0$ as

$$S_{p\upsilon}(f) = \frac{1}{T} P_0(f, T) V_0^*(f, T), \tag{38}$$

$$S_{\upsilon\upsilon}(f) = \frac{1}{T} V_0(f, T) V_0^*(f, T), \tag{39}$$

$$S_{pp}(f) = \frac{1}{T} P_0(f, T) P_0^*(f, T) \tag{40}$$

where superscript * denotes complex conjugate and T is the finite time of the record used in Fourier Transform as

$$P_0(f, T) = \frac{1}{T} \int_0^T P(0, t) \mathrm{e}^{-\imath\omega t} \mathrm{d}t \tag{41}$$

Therefore, taking FFT of Eqs. (35) and (36),

$$P_0(f, T) = A(f, T) + B(f, T) \tag{42}$$

$$V_0(f, T) = \frac{1}{Y} (A(f, T) - B(f, T)) \tag{43}$$

Substituting Eqs. (42) and (43) in Eqs. (38), (39) and (40) yields

$$S_{p\upsilon}(f) = \frac{1}{Y} \{S_{AA}(f) - S_{BB}(f) - 2\imath Q_{AB}(f)\} \tag{44}$$

$$S_{\upsilon\upsilon}(f) = \frac{1}{Y^2} \{S_{AA}(f) - S_{BB}(f) - 2C_{AB}(f)\} \tag{45}$$

$$S_{pp}(f) = S_{AA}(f) - S_{BB}(f) - 2C_{AB}(f) \tag{46}$$

where

$$S_{AA}(f) = \frac{1}{T} A(f) A^*(f) \tag{47}$$

$$S_{BB}(f) = \frac{1}{T} B(f) B^*(f) \tag{48}$$

$$S_{AB}(f) = \frac{1}{T} A(f) B^*(f) = C_{AB}(f) + \imath Q_{AB}(f) \tag{49}$$

Again, doing FFT (fast Fourier transform) of Eqs. (33) and (34) at $x = 1$ and $x = 2$,

Pressures,

$$P_1(f,T) = A(f,T)\mathrm{e}^{-\imath kx_1} + B(f,T)\mathrm{e}^{\imath kx_1} \tag{50}$$

$$P_2(f,T) = A(f,T)\mathrm{e}^{-\imath kx_2} + B(f,T)\mathrm{e}^{\imath kx_2} \tag{51}$$

and velocities,

$$V_1(f,T) = \frac{1}{Y}\left[A(f,T)\mathrm{e}^{-\imath kx_1} - B(f,T)\mathrm{e}^{\imath kx_1}\right] \tag{52}$$

$$V_2(f,t) = \frac{1}{Y}\left[A(f,T)e^{-\imath kx_2} - B(f,T)e^{\imath kx_2}\right] \tag{53}$$

Now,

$$\begin{aligned} S_{11}(f) &= \frac{1}{T}P_1(f,T)P_1^*(f,T) \\ &= S_{AA}(f) + S_{BB}(f) + 2[C_{AB}\cos(2kx_1) + Q_{AB}\sin(2kx_1)] \end{aligned} \tag{54}$$

$$\begin{aligned} S_{22}(f) &= \frac{1}{T}P_2(f,T)P_2^*(f,T) \\ &= S_{AA}(f) + S_{BB}(f) + 2[C_{AB}\cos(2kx_2) + Q_{AB}\sin(2kx_2)] \end{aligned} \tag{55}$$

$$S_{12}(f) = \frac{1}{T}P_1(f,T)P_2^*(f,T) = C_{12}(f) + \imath Q_{12}(f) \tag{56}$$

where

$$\begin{aligned} C_{12} &= S_{AA}\cos(k(x_1 - x_2)) + S_{BB}\cos(k(x_1 - x_2)) \\ &\quad + 2[C_{AB}\cos(k(x_1 + x_2)) + Q_{AB}\sin(k(x_1 + x_2))] \end{aligned} \tag{57}$$

$$Q_{12} = -S_{AA}\sin(k(x_1 - x_2)) + S_{BB}\sin(k(x_1 - x_2)) \tag{58}$$

If written in matrix form, where $s = x_1 - x_2$

$$\begin{bmatrix} S_{11} \\ S_{22} \\ C_{12} \\ Q_{12} \end{bmatrix} = \begin{bmatrix} 1 & 1 & 2\cos(2kx_1) & 2\sin(2kx_1) \\ 1 & 1 & 2\cos(2kx_2) & 2\sin(2kx_2) \\ \cos(ks) & \cos(ks) & 2\cos(k(x_1 + x_2)) & 2\sin(k(x_1 + x_2)) \\ -\sin(ks) & \sin(ks) & 0 & 0 \end{bmatrix} \begin{bmatrix} S_{AA} \\ S_{BB} \\ C_{AB} \\ Q_{AB} \end{bmatrix} \tag{59}$$

i.e.

$$[P_{12}] = [T][P_{AB}]$$

If taken inverse of [T], the values S_{11}, S_{22}, C_{12} and Q_{12} can be used to obtain the values of S_{AA}, S_{BB}, C_{AB} and Q_{AB} as follows:

$$[T]^{-1} = \frac{1}{6+2\cos(2k(x_1+x_2))} \begin{bmatrix} 1 & 1 & -2\cos(k(x_1-x_2)) & -\frac{3+\cos(2k(x_1+x_2))}{\sin(k(x_1-x_2))} \\ 1 & 1 & 2-2\cos(k(x_1-x_2)) & -\frac{3+\cos(2k(x_1+x_2))}{\sin(k(x_1-x_2))} \\ -\cos(2kx_2) & -\cos(2kx_1) & 2\cos(k(x_1+x_2)) & 0 \\ -\sin(2kx_2) & -\sin(2kx_1) & 2\sin(k(x_1+x_2)) & 0 \end{bmatrix} \tag{60}$$

$$[P_{AB}] = [T]^{-1}[P_{12}]$$

The above acquired matrix $[P_{AB}]$ can be used to calculate normalized impedance

$$\frac{Z(f)}{Y} = Z_n(f) = \frac{S_{AA}(f) - S_{BB}(f) - 2\iota Q_{AB}(f)}{S_{AA}(f) + S_{BB}(f) - 2C_{AB}(f)}$$

And the reflection coefficient

$$R(f) = |R|e^{\iota\theta} = \frac{Z_n(f) - 1}{Z_n(f) + 1} = \frac{-S_{BB}(f) + C_{AB}(f) - \iota Q_{AB}(f)}{S_{AA}(f) + C_{AB}(f) - \iota Q_{AB}(f)}$$

4 Measurement (Microphones and Transducers)

A microphone acts as a converter of acoustic property at the point of measurement to a more scalable quantity for computation, like voltage signal. The obtained voltage signal can be further used for analysis and computation. While selecting the type of microphone, one should pay special attention to the various specifications of the microphone and the experimental requirements like sensitivity, sampling rate, dynamic range (signal min-max), operation requirements (operating voltage and frequency), precision, accuracy and other parameters. The mismatch in any of these quantities may lead to inaccurate results; e.g. sampling rate should be at least twice the maximum measured frequency [6]; piezoelectric transducers may be used for robust applications; a high-voltage input may be required for high sensitivity measurements.

4.1 Sensitivity

Microphone sensitivity (S) is typically measured with a 1 kHz sine wave at a 94 dB sound pressure level (SPL), or 1 Pascal (Pa) pressure. The magnitude of the analog or digital output signal from the microphone with that input stimulus is a measure of its sensitivity. It is usually expressed as decibels (*dB*) with respect to a reference sensitivity of 1 V/Pa. Numerically [7],

$$S = 20 \times \log_{10} \frac{V \times p_0}{V_0 \times p} \text{dB} \quad \text{re} \quad \frac{1\,\text{V}}{\mu\text{Pa}}$$

Here, $p_0 = 1\,\mu\text{Pa}$ and $V_0 = 1\,\text{V}$.

4.2 Frequency Response

Frequency response is the magnitude and phase of the measured quantity as a function of frequency, with respect to input. For an acoustic system, the signal recorded should be undistorted, i.e. the frequency response line should be flat. But in practical case scenario, it is not so. Therefore, it needs to be studied for the region of concern and a particular microphone is selected accordingly.

Sample data for a typical microphone is provided in Fig. 6 [8].

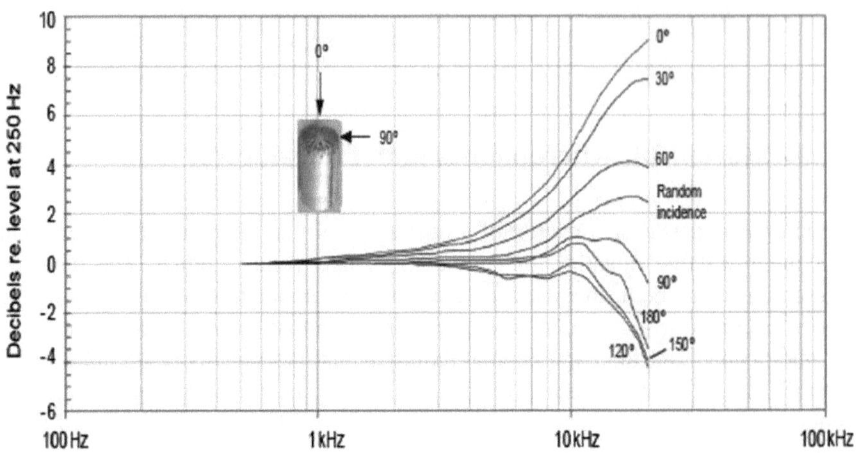

Fig. 6 Typical data for a 1/2″ microphone

4.3 Dynamic Range

It is the ratio of the largest to smallest value of the quantity being measured on decibel scale. A typical transducer has its own operating noise which distorts the very low values of the signal deciding the lower cut-off and very high values leading to saturation of the signal at peaks, thus deciding the higher cut-off.

Therefore, highly sensitive microphones are to be used for measuring very low signal values and low sensitivity microphones for high-level signals like impact noise.

5 Pre-amplification

The signal strength of microphone output is very small and if it is used to perform further processing, the SNR (signal-to-noise ratio) of that signal may get distorted. SNR is defined as the ratio of the strength of electrical signal output from microphone to that of unwanted interference. Mathematically,

$$\mathrm{SNR} = 10 \times \log_{10} \frac{P_{\mathrm{S}}}{P_{\mathrm{N}}}$$

where

P_{S} is the signal strength and
P_{N} is the noise strength.

As the noise is introduced mostly by the environmental disturbances and generally remains constant for a particular environment, P_{N} remains constant. The SNR can be improved simply by increasing P_{S}. So, pre-amplification is done to improve the SNR and thus accuracy of the measured signal. Due to this reason, the pre-amplifier is installed near the microphone, so as to minimize the distortion during transmission.

Most of the measurement devices come with inbuilt set-up for pre-amplification, and there is no need to add additional pre-amplifiers in the set-up.

6 Analog to Digital Conversion

As most of our processing and simulation is done through computers, the analog signal recorded by microphones is converted to digital signal via analog to digital converters. This signal is fed to the computer for further calculations and generation of the results.

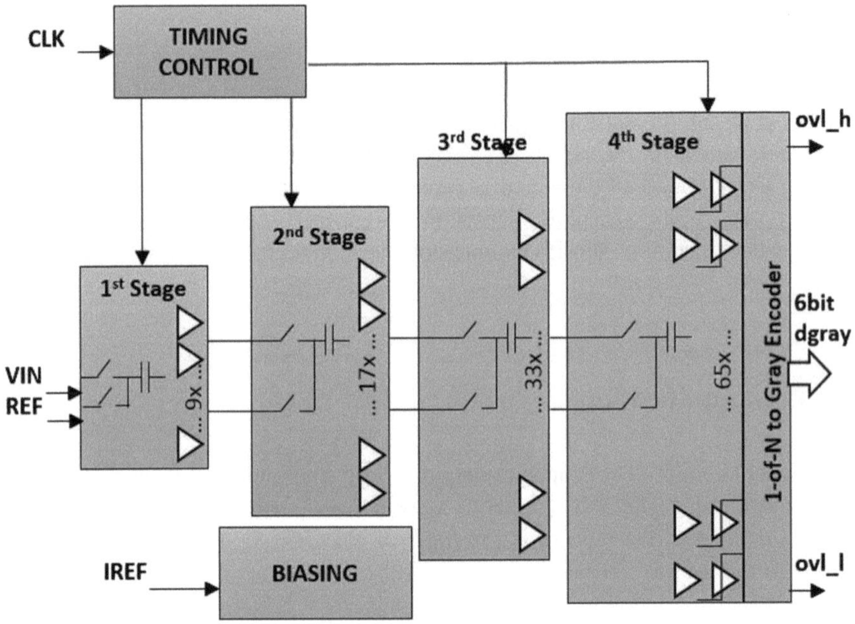

Fig. 7 Flash ADC

There are many methods for conversion of analog signal to digital signal given below.

6.1 *Flash ADC*

It is mostly used, where high sampling rate with low to moderate resolution is required. But high power consumption, size and cost are associated for higher precisions. A typical ADC is shown in Fig. 7 [9].

6.2 *Sigma-Delta ADC*

It encodes the analog signal to digital with at higher frequency, then applies digital filter to form high-resolution low sampling frequency digital output as shown in Fig. 8 [10]. It provides high resolution without the need of external precision equipment, but is slower due to oversampling.

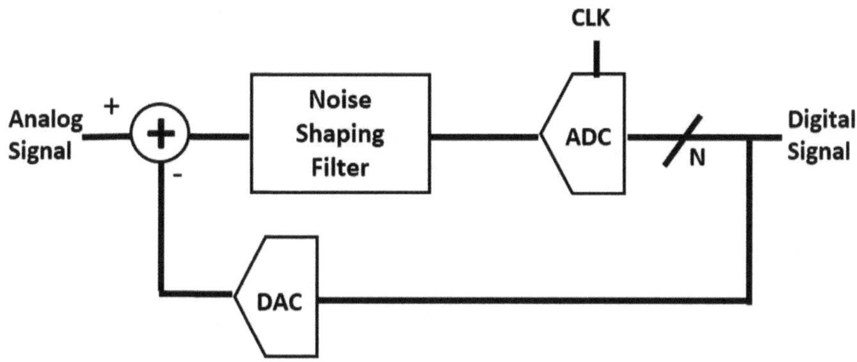

Fig. 8 Sigma-delta ADC

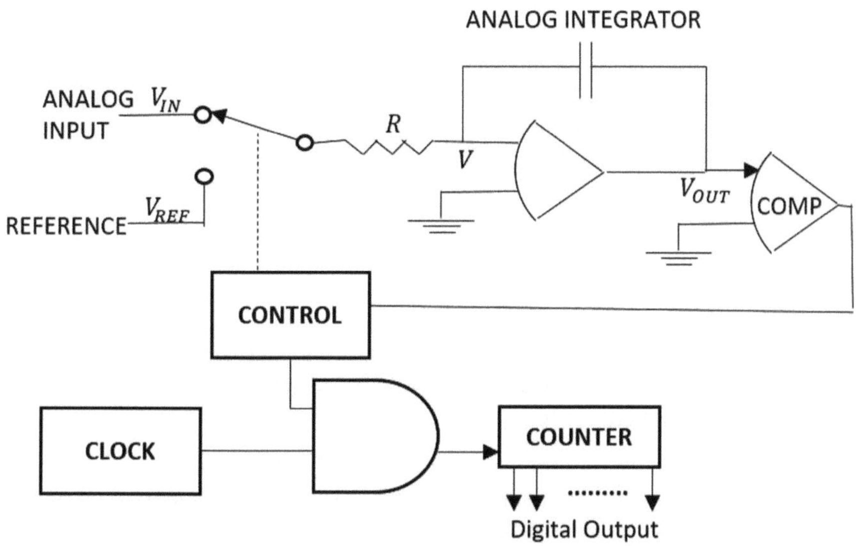

Fig. 9 Dual slope ADC

6.3 *Dual Slope ADC*

The sampled signal charges the capacitor for a fixed amount of time [11]. By integrating over time, noise integrates out of the conversion. Then, the ADC discharges the capacitor for a fixed rate, while a counter counts the ADCs output bits (Fig. 9). A longer discharge time results in a higher count. The signal is more immune to noise and has high accuracy, but it is slow and requires high precision external components to achieve accuracy.

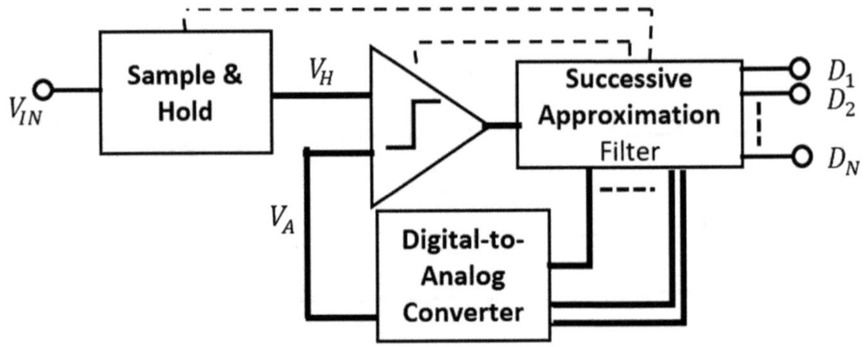

Fig. 10 Successive approximation converter

6.4 Successive Approximation Converter

It sets MSB (most significant bit) and converts it to analog form using DAC [12]. Then, it compares guess to input, sets next MSB and tests again. It is capable of high speed and lower cost, but has medium accuracy. Also, the speed is limited to 5 Msps and high resolution leads to slower speed. A sample SAC is shown in Fig. 10.

7 Analysis

The obtained signal in digital form is fed to the computer, and the formulation is done according to the above theory. The calculation and result generation can be done using various methods like

- Direct feeding to the custom designed circuits which use various arithmetic and logic gate circuits, and measuring the output signal for obtaining the required results. A circuit can be generated using operational amplifiers.
- Coding on software like MATLAB so that on feeding the digital signal in matrix form, it generates result.
- Making a block diagram using signal processing approach for the required framework calculations on software like LABVIEW.

8 Precautions

While performing an analysis, it is important to know what is to be done in the experiment, but it is equally and sometimes more important to know the things which are not to be done or taken care of, during the analysis. A little negligence in the procedure may lead to drastically erroneous results, which may become useless even after such an elaborate set-up and hard work. Some of them include the following:

8.1 Deviation from Assumptions

In the current analysis, there is an assumption that the travelling wave is planar in nature. But the acoustic waves remain planar in a duct, only up to a specific upper bound frequency. Above this frequency, the wave becomes non-planar and the results tend to include much higher noise content.

The planar wave frequency [2] for a cylindrical duct waveguide of diameter d in metre is

$$f_{\text{planewave}} \leq \frac{202}{d}$$

Taking two different frequencies for a given diameter pipe, one in the planar wave region and the other in the non-planar wave region, the following output has very high noise content for deviation from the assumption.

As seen in the graphs in the end (Figs. 11 and 12), as the frequency is increased above the given limit, the non-planar wave is generated. As the theory above is formulated with respect to planar waves, the non-planar part is treated as noise. The resulting analysis is full of noise. As seen in the graphs, the planar wave gives much smoother curve than the non-planar wave, when plotted for wave number for the same tube.

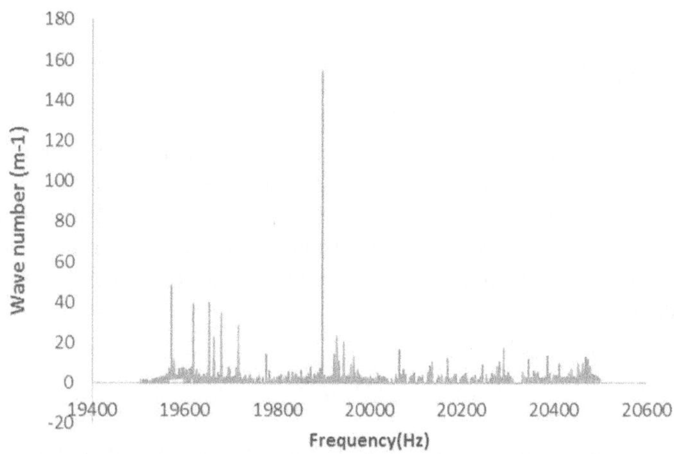

Fig. 11 Wave number for frequency corresponding to non-planar waveform

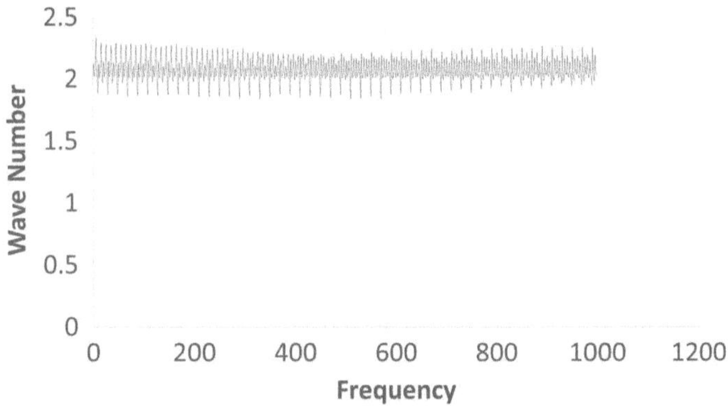

Fig. 12 Wave number for frequency corresponding to planar waveform

8.2 *Environmental Effects*

Always keep track of all the variables involved in the analysis, which tend to modify with the variation of environmental factors. One may calculate a specific result in a particular scenario or use it from some other source, but environmental conditions involved in the analysis of the source are to be taken care of. In our case, the speed of sound c_0 depends on surrounding temperature and altitude [13].

$$c_0 = \sqrt{\frac{\gamma R_0 T}{m_0\left(1 + \frac{\gamma m_0 g^2 z^2}{C_p R_0 T}\right)}}$$

Some sample data values are provided in Table 1 to show the variation of the acoustic wave velocity with respect to the environmental factors.

8.3 *Special Anomalies*

Sometimes, the property under test involves some special or exceptional configurations which may produce some strange results, even when the experimental conditions are well taken care of. A similar special case scenario is experienced when the distance between the two microphones reaches a particular value. The reflection coefficient becomes indeterminate at that particular frequency [14]. Mathematically,

$$s \leq \frac{c_0}{2 f_{\text{mean}}}$$

It can be seen in the graphs plotted (Figs. 13 and 14) that there is a huge discontinuity in the curve at about 2800 Hz. This is obtained due to microphone

Table 1 Acoustic wave velocity

Atmospheric temperature T(K)	Altitude z(m)	Acoustic wave velocity c_0(m/s)
288.15	0	340.3
287.82	50	340.11
287.5	100	339.91
286.85	200	339.5
286.2	300	339.06
285.55	400	338.61
284.9	500	338.13
284.25	600	337.64
283.6	700	337.12
282.95	800	336.59
282.3	900	336.03
281.65	1000	335.45
275.15	2000	328.47
268.65	3000	319.34
262.15	4000	308.12
255.65	5000	295.04
249.15	6000	280.42
242.65	7000	264.66
236.15	8000	248.18
229.65	9000	231.38
223.15	10,000	214.65
216.65	11,000	198.26

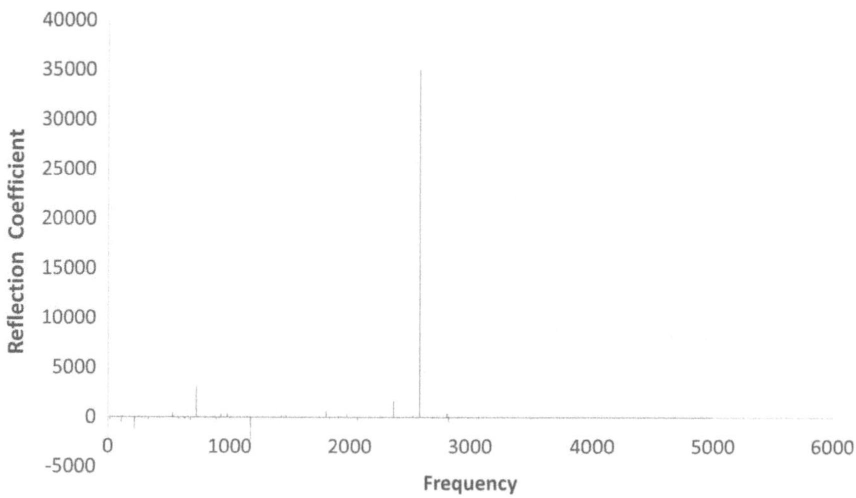

Fig. 13 Reflection coefficient when microphone spacing is integral multiple of half wavelength of test signal

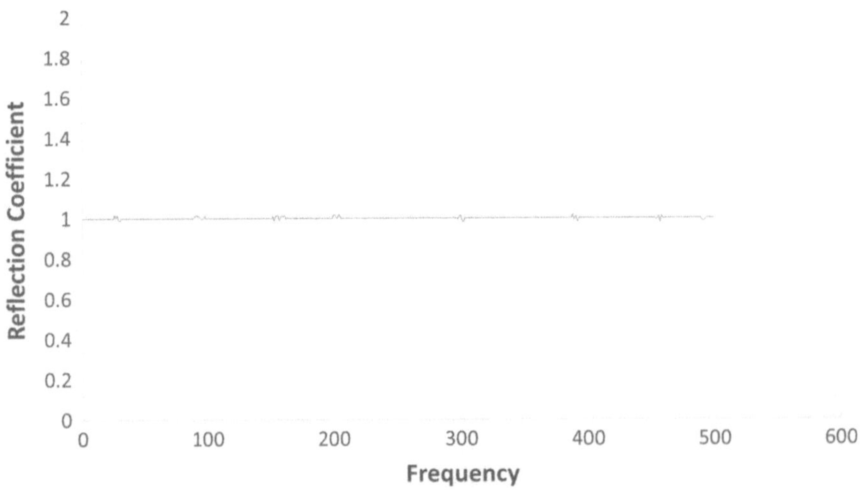

Fig. 14 Reflection coefficient when microphone spacing is less than half wavelength of test signal

spacing of 6 cm. If the lower frequencies are used or the microphone spacing is reduced, the problem can be completely eliminated.

8.4 *Methods to Enhance Accuracy*

Most of the assumptions are made on the basis of approximation and at the cost of accuracy. These approximations are made to reduce the computational complexity and sometimes additional set-up requirements for the required analysis. In our case, as a transient signal is being recorded, there is an instrument phase mismatch [15] introduced while acquiring the data from two different sources. There is no additional calculations done while formulation, to eliminate this error. But this error can be eliminated by introducing a little modification in the transfer function between the two microphones. This transfer function is proportional to cross-spectral density. So, the method suggests to use the original and switched microphone configurations to generate the two different transfer functions. The geometric mean of these transfer functions is the original transfer function required with instrument phase mismatch eliminated. Mathematically,

$$H_{12} = \sqrt{H_{12}^{\text{original}} H_{12}^{\text{switched}}}$$

where

$$H_{12} = \frac{S_{12}}{S_{11}}.$$

References

1. Munjal M (2014) Acoustics of ducts and mufflers, 2nd edn. Wiley, Chichester
2. Russell DA (2004) Absorption coefficients and impedance. Science and Mathematics, Department, GMI Engineering & Management Institute, www.gmi.edu, diakses Maret 1–6
3. Lima KF, Barbieri N, Terashima FJ, Rosa VP, Barbieri R (2015) Preliminary evaluation of the sound absorption coefficient of a thin coconut coir ber panel for automotive applications. J Acoust Soc Am 138(3):1887–1887
4. Song BH, Bolton JS (2000) A transfer-matrix approach for estimating the characteristic impedance and wave numbers of limp and rigid porous materials. J Acoust Soc Am 107(3): 1131
5. Cerna M, Harvey AF (2000) The fundamentals of FFT-based signal analysis and measurement in LabVIEW and LabWindows/CVI. Spectrum 41(July):1–20
6. Lerman G, The Shannon sampling theorem and its implications, vol 1, pp 1–10
7. Malchaire J, Sound measuring instruments [online]. http://www.who.int/occupational_health/publications/noise6.pdf
8. G.R.A.S, Product catalogue - measurement microphones. Sound & Vibration [online]. http://www.gras.dk/media/MiscFiles/SalesMat/GRASProductCatalogue_WEB.pdf
9. Cmos D, Sandner C, Clara M, Santner A, Hartig T, Kuttner F (2005) A 6-bit 1.2-GS/s low-power flash-ADC in 0.13-"m Digital CMOS Christoph. IEEE J Solid-State Circuits 40 (7):1499–1505
10. Miller MR, Petrie CS (2003) A multibit sigma delta ADC for multimode receivers. IEEE J Solid-State Circuits 38(3):475–482
11. Mutoh A and Nitta S (1999) Noise immunity characteristics of dual-slope integrating analog-digital converters. In: 1999 international symposium electromagnetic compatibility, pp 622–625
12. Sauerbrey J, Schmitt-Landsiedel D, Thewes R (2003) A 0.5-v 1-μW successive approximation adc. IEEE J Solid-State Circuits 38(7):1261–1265
13. Kirtskhalia VG (2012) Speed of sound in atmosphere of the earth. Open J Acoust (June), 80–85
14. Chung JY, Blaser D (1980) Transfer function method of measuring in-duct acoustic properties. Acoust Soc Am 68(3):907–921
15. Chung JY (1978) Cross-spectral method of measuring acoustic intensity without error caused by instrument phase mismatch. J Acoust Soc Am 6:1613

Optimal Inspection Intervals for Multi-One-Shot Systems

Li Liu, Young Jin Han, Ha Won Kim and Won Young Yun

1 Introduction

One-shot systems, such as missile systems, munitions, fire extinguishers, and automobile air bags are usually kept in storage and taken into use when required. It is important to keep system reliability in a high level during storage in order to ensure the operation can be performed successfully. But, the reliability of system in storage deteriorates with time, and the exact failure time of system cannot be observed. We can only observe the condition of system either it is taken into operation or by inspection. Thus, inspection must be carried out periodically to detect system failure immediately and keep system reliability in a high level. Excessive inspection results in a reduction in the number of defective units in the system but high maintenance costs. While, if inspections are performed less frequently to reduce maintenance costs or preserve the electronic components, system reliability will be decreased. Briefly, an appropriate inspection and maintenance policy improves system reliability, decreases maintenance costs, and lengthens units' lifetime. Thus, inspection intervals of one-shot systems which represent an appropriate balance between maintenance costs and system reliability should be determined optimally. Usually, one-shot systems consist of many units with a series structure and the units can be classified into two types where Type 1 units fail at random times and Type 2 units are degraded with time. For example, in the case of a missile system the guidance and control units are Type 1 units, while the ignition unit of the rocket motor is a Type 2 unit. Especially, Type 2 units require stringent storage environments in order to prevent its degradation, and in some cases, the

L. Liu · H.W. Kim · W.Y. Yun (✉)
Department of Industrial Engineering, Pusan National University, Busan, South Korea
e-mail: wonyun@pusan.ac.kr

Y.J. Han
Reliability Engineering Team, Hyosung Corporation, Chang-won, South Korea

M. Ram and J.P. Davim (eds.), *Modeling and Simulation in Industrial Engineering*, Management and Industrial Engineering, DOI 10.1007/978-3-319-60432-9_3

process of testing may degrade the electronic components of units in the system. Hence, it is also necessary to determine the preventive replacement ages of Type 1 units and preventive maintenance thresholds of Type 2 units optimally. Nakagawa and Mizutani [1] and Nakagawa et al. [2] reviewed the maintenance and inspection policies over a finite time span. Ito and Nakagawa [3] dealt with a periodic inspection policy problem for a storage system which is composed of two types of units, where Type 1 units are maintained upon inspection, while Type 2 units are degraded with time. Also, they considered an inspection policy in which the system is replaced if its reliability becomes lower than a predetermined level, and the optimal inspection time, T^*, which minimizes the average cost until overhaul, was determined. Kaio et al. [4] dealt with an inspection policy problem for a single-unit system and determined the optimal inspection times which minimize the expected cost until the detection of the system failure. Ito and Nakagawa [5] considered a storage system consisting of three types of units where Unit 1 is maintained at time interval T, Unit 2 is not maintained but replaced at time interval NT, and Unit 3 is neither maintained nor replaced. They determine the optimal inspection time, T^*, and optimal inspection number, N^*, which minimize the average total cost until overhaul. Also, Ito and Nakagawa [6] dealt with the different inspection policy in which the system is replaced at the detection of failure or at time $(N + 1)T$, whichever occurs first. The optimal inspection time, T^*, which minimizes the total expected cost until replacement, was determined. Under the inspection policy proposed by Ito and Nakagawa [3], Ito and Nakagawa [7] determined the optimal inspection time which minimizes the total expected cost, until the detection of the system failure. Later Ito and Nakagawa [8] considered different costs as testing cost and overhaul cost and determined the optimal inspection times which minimize the average cost until overhaul. Taghipour and Banjevic [9] proposed an inspection policy in which the failed units are either minimally repaired or replaced at predetermined times, as a function of their age. The optimal inspection interval, T^*, was determined to minimize the expected total cost. The inspection intervals may be reduced as the system age increases, thus the regular inspection intervals may not be always profitable. Golmakani and Moakedi [10] dealt with a non-periodic inspection policy for a multi-unit system and considered that the failure of the units does not cause the system but incurs a penalty cost due to the time delay between failure and detection. They compared the non-periodic inspection policy with a periodic inspection policy.

In some cases, there may have no knowledge about the lifetime distribution of the system and the inspection may not be perfect in the sense that it cannot reveal the true state of the system without error. Leung [11] dealt with an imperfect inspection problem for a single-unit system with the unknown lifetime distribution and considered the inspection models with type I and II error. For complex repairable systems, system availability is the more appropriate performance measure and Cui and Xie [12] studied the availability of a storage system with an periodic inspection policy and proposed two periodic inspection policies where one is that the condition of the system is as good as new after performing inspection, and the other one is that the condition of the system is as bad as old before performing inspection. Kitagawa et al. [13] studied a periodic inspection policy of a one-shot systems with multi units. They considered minimal repair when failures are detected by inspection and when the nth failure is detected, and all units are replaced by new ones. Kitagawa et al. [14] proposed periodic and non-periodic inspection policies for a one-shot system under minimal repair and determined the optimal number of failures before replacement and the inspection intervals to minimize the expected cost per unit time ensuring a predetermined mean availability. Yun et al. [15] studied an inspection policy problem for a one-shot system with two types of units over a finite time span, where Type 2 units should be replaced by new ones at predetermined times. The interval availability betweeninspections was used as an optimization criterion, and a simulation-based optimization procedure with a genetic algorithm was proposed to find near-optimal inspection intervals that minimize the life cycle cost and satisfy the target interval availability.

For evaluating the performance of one-shot systems accurately, a mathematical model is required to describe the degradation of Type 2 units in one-shot systems and stochastic processes, such as Brownian motions, have often been used. Grall et al. [16] considered the single-deteriorating system with a gamma process and proposed a condition-based maintenance policy. Optimal inspection times and preventive maintenance thresholds were determined, and also a periodic inspection scheme was compared with a dynamic inspection scheme. Avramidis et al. [17] explained the procedure of the gamma-bridge sampling algorithm in detail. Li and Pham [18] proposed a condition-based maintenance model for a storage system and proposed two randomized degradation functions to describe the degradation of the units in the system. The inspection intervals and preventive maintenance thresholds of the units that minimize the expected total maintenance cost were determined. Noortwijk [19] reviewed the applications of gamma processes in time-based and condition-based maintenance optimization problems, and explained two algorithms, gamma-increment sampling and gamma-bridge sampling algorithms, to simulate gamma processes. Yun et al. [20] dealt with an inspection policy problem for a one-shot system and the compound Poisson process for the degradation process of Type 2 units was considered. Later, Yun et al. [21] considered a gamma process to model the degradation process of Type 2 units and apply a gamma-bridge sampling method for simulating the gamma process. A genetic algorithm was proposed to find near-optimal inspection intervals and preventive maintenance thresholds of Type 2 units that minimize the life cycle cost and satisfy the target interval availability.

In practice, there are usually many one-shot systems in storage and maintenance sites have a certain amount of maintenance resources such as engineers, maintenance equipment, and spare parts. Thus, the maintenance delay time due to shortage of maintenance resources at maintenance sites should be considered when the inspection intervals are determined. Yun et al. [22] studied an inspection policy for one-shot systems under limited maintenance resources. They determined the optimal starting point of each system, optimal inspection intervals, and the optimal preventive threshold of Type 2 units to minimize the life cycle cost and satisfy the target interval availability by a genetic algorithm. Yun et al. [23] dealt with an inspection scheduling problem for one-shot systems under limited maintenance resources. They considered the degradation of Type 2 units, and used two randomized degradation functions proposed by Li and Pham [18], as well as a gamma process, to describe the degradation of Type 2 units. A genetic algorithm was proposed to determine near-optimal inspection intervals, preventive replacement age, preventive maintenance threshold, and starting point of inspection for one-shot systems, with the aim of minimizing life cycle cost and satisfying target interval availability. However, they did not evaluate the performance of the proposed genetic algorithm.

In this chapter, we deal with an inspection schedule problem for multi one-shot systems which are composed of two types of units. The interval availability and life cycle cost are used as optimization criteria and are estimated by simulation. For the optimization of the inspection schedule of one-shot systems, we aim to determine the inspection intervals, preventive replacement ages of Type 1 units, and preventive maintenance thresholds of Type 2 units, as well as the starting point of inspection for each one-shot system optimally. This chapter is organized as follows. The inspection schedule model of one-shot systems is introduced in Sect. 2. Next, a simulation-based optimization procedure with two phases is proposed in Sect. 3 and numerical examples are studied to compare the heuristic method with the existing genetic algorithm in Sect. 4. Finally, Sect. 5 concludes the study.

2 Inspection Schedule for Multi One-Shot Systems Under Limited Maintenance Resources

In this section, we introduce an inspection schedule model for multi one-shot systems with two types of units and explain the appropriate performance measures as interval availability and life cycle cost, to operate one-shot systems effectively. Table 1 shows notation for the inspection schedule model of one-shot systems.

2.1 Inspection Schedule for Multi One-Shot Systems

In practice, there generally exists a maintenance system with a multi-echelon structure as operational sites, inspection sites, and depots for the inspection and maintenance of one-shot systems [15]. At the beginning of the operational phase, one-shot systems are deployed in the operational sites and are then stored for a long period of time until their inspection or operation. In order to carry out the inspection of one-shot systems, the systems are moved to the maintenance sites and their inspection and maintenance are performed at inspection sites or depots based on their failure types. Usually, one-shot systems are first moved from operational sites

Table 1 Notation in the inspection schedule model of one-shot systems

Notation	Definition
i	Index of one-shot systems (i = 1, 2, 3, … , N)
j	Index of units (j = 1, 2, 3, … , M)
p	Index of periods (p = 1, 2, 3, … P)
ε	Allowable gap
θ	Uniform random variable ($0 < \theta < 1$)
Mu	Mutation rate
AI_p	Interval availability of pth period
AI_T	Target interval availability
A_{sys}	System availability
C_S	Set-up cost
C_I	Inspection operation cost
C_{CM}^j	Corrective maintenance cost of unit j
C_{PM}^i	Preventive maintenance cost of unit j
LC	Life cycle cost
TI	Total number of inspection
TI_i	Total number of inspection of system i
TCM_i^j	Total number of corrective maintenance of unit j in system i
TPM_i^j	Total number of preventive maintenance of unit j in system i
A_{STD}	Standard deviation of interval availability
A_{STD_T}	Target standard deviation of interval availability
ΔA_{STD}	Increment in system availability
ΔCost	Increment in life cycle cost
IP_{Opt_Fisrt}	Optimal first inspection point determined in Phase 1
IP_{Opt_Second}	Optimal second inspection point determined in Phase 1
$IP_{Earliest}$	Earliest first inspection time ($0 < IP_{Earlies} \leq IP_{Opt\_Fisrt}$)
IP_{Latest}	Latest first inspection time($IP_{Opt\_Fisrt} \leq IP_{Latest} \leq IP_{Opt\_Second}$)

to inspection sites in order to check whether they can operate normally or not. If the system failure is not detected at the inspection site, they are returned to the operational site. Otherwise, the failed systems are moved to depots for the replacement or repair of the failed units in the systems, and the inspection is carried out again until the condition of the systems becomes good. On the other hand, each maintenance site necessarily contains a certain amount of maintenance resources, such as spare parts, repairmen, and maintenance equipment. If all one-shot systems are inspected and maintained at the same time, the maintenance delay time due to shortage of maintenance resources increases. As a result, the performance measure (availability) of one-shot systems at the operational site may decrease. Thus, we need to establish the inspection schedule of one-shot systems under the constraint of maintenance resources. For example, we consider 3 one-shot systems that are inspected at every 8 years and Fig. 1-(I) shows that all one-shot systems are

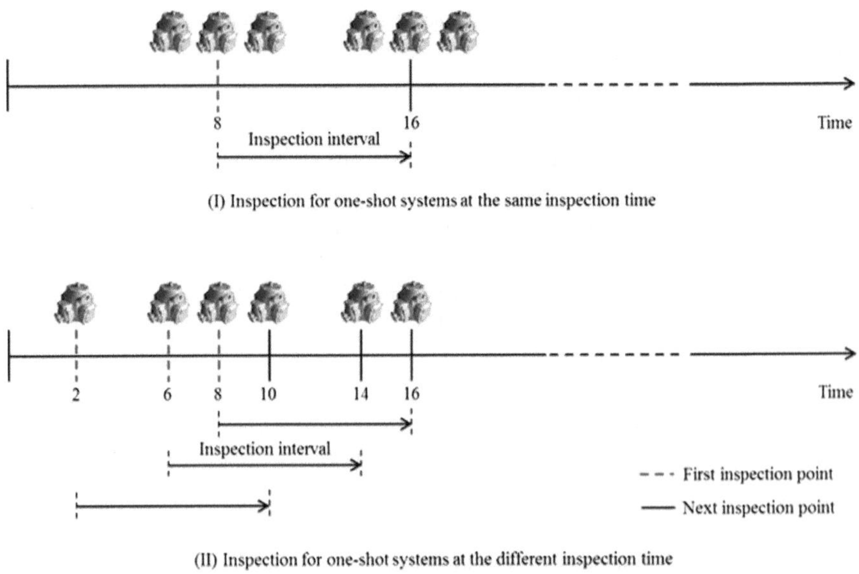

Fig. 1 An example of the starting points of inspection

inspected simultaneously at every inspection period. If there is a limitation of maintenance resources in the inspection site and depot, one-shot systems will take much more time to go back to the operational site. While Fig. 1-(II) shows that each one-shot system may be inspected at different times, where the first inspection point of each one-shot system differs. The maintenance delay time incurred through this method will be less than that incurred where the inspection schedule of each system is the same, but it may incur higher inspection operation costs due to increasing the number of inspections. Thus, we aim to determine the optimal inspection intervals and starting points of inspection for one-shot systems in order to optimize the inspection schedule of one-shot systems.

Usually, one-shot systems consist of two types of units with a series structure, where Type 1 units suffer increasing wear with their age and may fail at random times. In one-shot systems, each Type 1 unit may have a different failure distribution including an exponential distribution or a Weibull distribution. If the time to failure of the Type 1 units follows an exponential distribution, there is no reason to replace Type 1 units preventively upon inspection [4]. While if the failure time of the Type 1 units follows a Weibull distribution with increasing failure rate, the Type 1 units are deteriorated and a risk of failure is increased with time. Therefore, we consider a preventive maintenance policy for the deteriorated Type 1 units in which the Type 1 units whose age reaches the preventive replacement age are replaced by new ones at inspection. On the other hand, Type 2 units are degraded with time and the Type 2 units whose degradation level exceeds the failure threshold cannot operate normally. For preventing the failure of Type 2 units, the Type 2 units whose degradation level reaches the preventive maintenance threshold of Type 2 units are

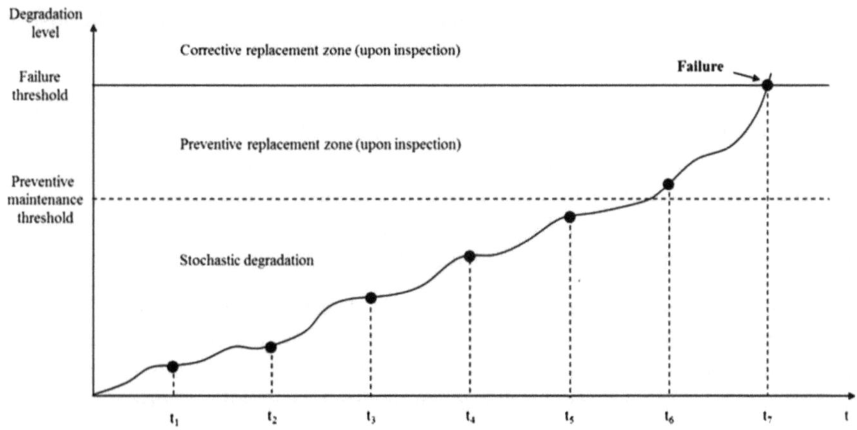

Fig. 2 An example of the degradation process of Type 2 units

replaced by new ones preventively as shown in Fig. 2 [8]. Therefore, we also need to determine the preventive maintenance threshold of Type 2 units in order to keep high system reliability in storage. Thus, we need to determine the preventive replacement age of Type 1 units and preventive maintenance threshold of Type 2 units optimally.

2.2 *Inspection Schedule Model of Multi One-Shot Systems*

In general, the steady state (long-run average) availability, which is interpreted as the average proportion of the working period of systems, is usually used as an optimization criterion for optimizing inspection and maintenance policies of complex repairable systems, such as railway systems, aircraft, and automobiles. However, it is not appropriate to evaluate the performance of one-shot systems, because a one-shot system is required to perform a mission suddenly between inspection times and only once during its life cycle. For such systems, keeping high system availability during inspection periods is important for performing mission successfully. Hence, we consider interval availability, which is defined by the fraction of time during which a system is in operation over a finite inspection period, as an optimization criterion [21]. According to the state of Type 1 and 2 units in one-shot system upon inspection, two kinds of maintenance action can be performed, namely, corrective maintenance (CM) and preventive maintenance (PM). Therefore the inspection of one-shot system can incur a series of cost including set-up cost, inspection operation cost, and CM and PM costs of Type 1 and 2 units during the system life cycle. Hence, the life cycle cost is considered as another optimization criterion to assess the system performance. In this chapter, we deal with an inspection schedule problem for one-shot systems with two types of units and consider the limited maintenance resources. Assumptions of the inspection schedule model for one-shot systems are as follows:

(1) The one-shot system consists of two types of units with a series structure.
(2) The failure of all units in the system is independent.
(3) The life cycle time of one-shot system is finite and given.
(4) The failure of a one-shot system is detected only by inspection.
(5) The failure threshold of Type 2 units is given.
(6) The perfect repair model is used for CM and PM.

Given the above assumptions, we want to determine the inspection intervals, starting point of the first inspection of each system, preventive replacement age of Type 1 units and preventive maintenance threshold of Type 2 units that minimize the life cycle cost and satisfy the target interval availability during inspection periods. The optimization problem is formulated as

$$\mathrm{MinLC} = (C_{\mathrm{S}} \times TI_i) + \sum_{i=1}^{N}\sum_{j=1}^{M}\left\{\left(C_{\mathrm{CM}}^{j} \times E\left[\mathrm{CM}_i^j\right]\right) + \left(C_{\mathrm{PM}}^{j} \times E\left[\mathrm{PM}_i^j\right]\right)\right\} \tag{1}$$

subject to $\mathrm{AI}_p \geq \mathrm{AI}_{\mathrm{T}}, \forall p$

3 Optimization Method for Inspection Schedule of Multi One-Shot Systems

In the previous section, we introduced the inspection scheduling model for one-shot systems under the constraint of maintenance resources and explained the decision variables as the inspection intervals, preventive maintenance ages, preventive maintenance thresholds, and starting points. In this section, we propose a method for optimizing the inspection schedule of multi one-shot systems. In this section, Firstly, a simulation-based optimization procedure with two phases is proposed to find near-optimal solutions that minimize the life cycle cost and satisfy the target interval availability over all periods. Next, we explain the basic frame of the simulation model to estimate the life cycle cost and interval availability of multi one-shot systems. Also, a general genetic algorithm and a hybrid genetic algorithm with a heuristic method are proposed in order to generate alternatives for the inspection schedule of multi one-shot systems.

3.1 Simulation-Based Optimization Procedure with a Hybrid Genetic Algorithm

In order to find the optimal inspection schedules for one-shot systems, we should firstly calculate the interval availability over all periods and life cycle cost

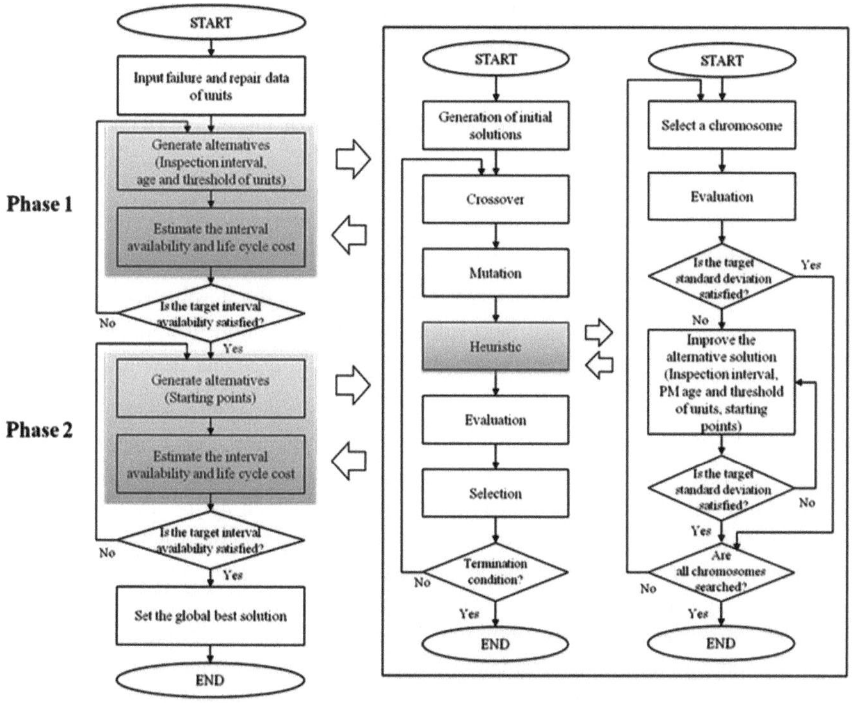

Fig. 3 Simulation-based optimization procedure with a hybrid genetic algorithm

accurately. However, it is difficult to calculate these analytically because each Type 1 unit may have a different failure distribution and the failure and repair of the units at the next inspection time depends on the ones at the previous inspection interval. Thus, we use simulation method to estimate the interval availability and life cycle cost. Additionally, appropriate optimization methods are required in order to find near-optimal alternatives, and we propose a general genetic algorithm and a hybrid genetic algorithm with a heuristic method. Thus, a simulation-based optimization procedure with a hybrid genetic algorithm is proposed for optimizing the inspection schedule of one-shot systems as shown in Fig. 3 and the detailed procedure is as follows:

Step 1 Input simulation data, such as failure and repair distributions of units, maintenance costs.

Step 2 Set the target interval availability.

Step 3 Generate alternatives of the inspection intervals, preventive replacement ages of Type 1 units and preventive maintenance thresholds of Type 2 units in a one-shot system through the hybrid genetic algorithm with the heuristic method in Phase 1.

Step 4 Estimate the interval availability over all periods and life cycle cost by simulation.

4.1 If $AI_p \geq AI_T$ (for all p), set that the current best solution is the global best solution in Phase 1 and go to Step 5. Otherwise, go to Step 3 to generate new alternatives.

Step 5 Generate alternatives for the starting point of inspection of each one-shot system through the hybrid genetic algorithm with the heuristic method in Phase 2.

Step 6 Estimate the interval availability over all periods and life cycle cost by simulation.

6.1 If $AI_p \geq AI_T$ (for all p), set that the current best solution is the global best solution in Phase 2 and terminate the procedure. Otherwise, go to Step 5 to generate new alternatives.

We used a simulation model that is based on the discrete-event method using a next-event time advance technique for advancing the simulation time [25]. The simulation model contains several types of processes such as the simulation main module, initialization module, timing module, and event controller module. The simulation main module controls the simulation behaviors and calls the initialization, timing, and event controller modules. The simulation begins at time '0', and the states of the system and components are initialized by the initialization routine. Next, the timing module determines the next-event type such as the failures of the components, inspection of the systems, and advances the simulation time. The event controller gets an event from the event list and invokes the event according to the type of the object (system and component) as shown in Fig. 4.

We consider the object-oriented design method and hence the simulation operation logic designs based on the activities of the objects which are regarded as key elements in modeling. The main two objects are defined as system and component, and their events that may change the state of each object are classified into global and interactive events as shown in Table 2. Global events are those which are enrolled into the event list by system or component, and interactive events are invoked by other objects. The objects have three states: ***RUN***, ***DOWN***, and ***IDLE***. These states will be changed when the simulation time reaches a specified time that invokes an event.

Figure 5 shows the state transition diagram of the system and component according to the event of each object. When the simulation starts or system operation begins again, the state of the system changes from ***IDLE*** to ***RUN*** and the system activates the components that the states of the components change from ***IDLE*** to ***RUN***. When ***BreakDown*** event of a component occurs, the state of the component changes from ***RUN*** to ***DOWN***. The state of the system will be changed from ***RUN*** to ***DOWN*** according to the reliability structures between the components. If the system fails, the ***deActivate*** event occurs in order to change the states of other components from ***RUN*** to ***IDLE***. ***EndFix*** event of the failed component can only be invoked when the system is inspected, and its state changes from ***DOWN*** to

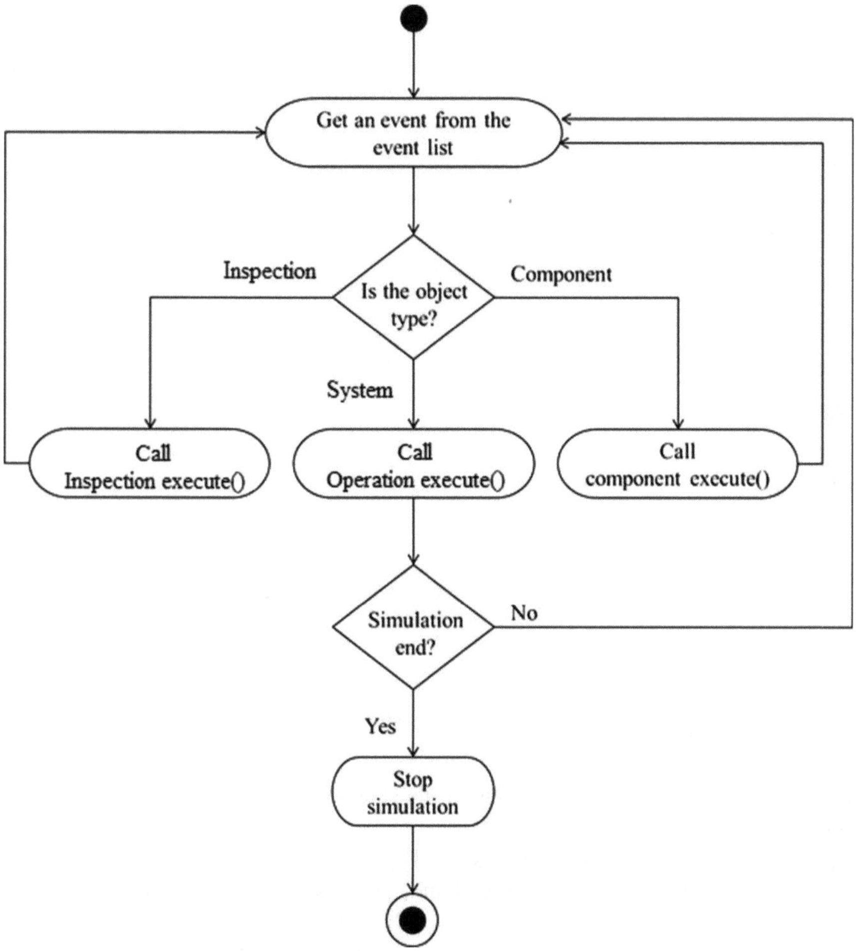

Fig. 4 Activity diagram of the event controller

Table 2 Events of system and component

Object	Type	Event	Description
System	Global	START	When the system operates
		STOP	When the system is inspected or failed
		END	When the simulation end time reaches
	Interactive	Activate	Activate components
		deActivate	Deactivate components
Component	Global	BreakDown	When the failure of a component occurs
		EndFix	When the failed component is fixed
	Interactive	Activate	When the system operates or begins again
		deActivate	When the system operates or is inspected

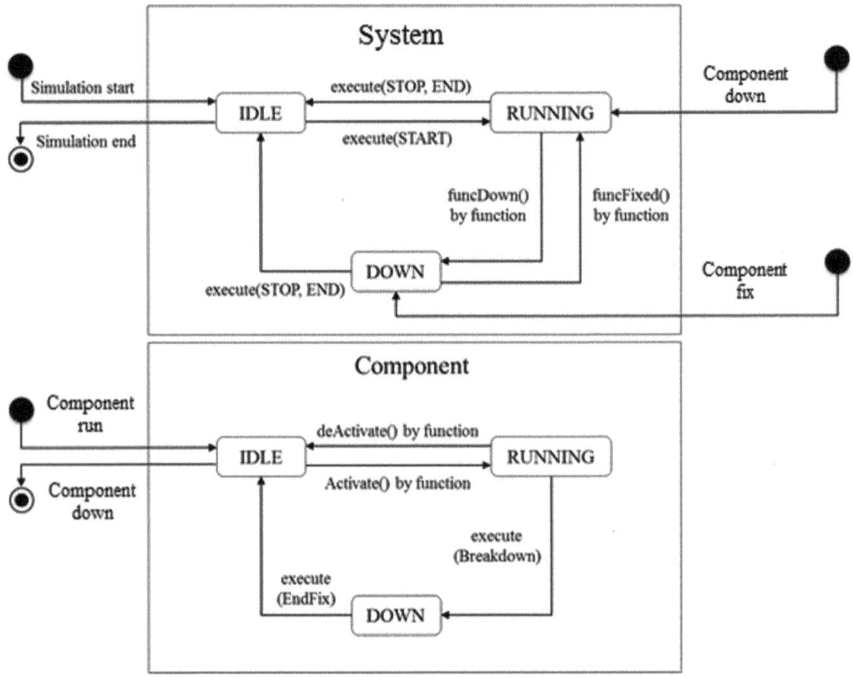

Fig. 5 State transition diagrams for system and component

IDLE. If the system can operate again, its state changes from ***IDLE*** to ***RUN*** and activates all components in the system.

For generating the failure times of Type 1 components, we can apply the well-known methods. On the other hand, stochastic processes such as Markov processes that include Brownian motion with drift, compound Poisson process, and gamma processes have been used for describing the degradation processes of Type 2 units [19]. Especially, gamma processes have been often used for the stochastic modeling of monotonic and gradual deterioration processes. Hence, we consider the gamma process to model the degradation processes of Type 2 units and apply a gamma-bridge sampling method for simulating the gamma process [17].

3.2 Hybrid Genetic Algorithm with Heuristic Method in Phase 1

In Phase 1, we consider a one-shot system and propose a hybrid genetic algorithm with heuristic method to find the inspection intervals, preventive replacement ages of Type 1 units, and preventive maintenance thresholds of Type 2 units that minimize the expected life cycle cost and satisfy the target interval availability [21].

In the genetic algorithm, a gene in a chromosome represents inspection year and 1.0 year is used as the unit time, the value of each gene is either ‘0’ or ‘1’. The integer strings are used to represent the preventive replacement ages of Type 1 units and preventive maintenance thresholds of Type 2 units. Figure 6 shows an example of the solution representation, and the one-shot system is inspected at 4, 12, 14, 26, and 29 years. Also, the Type 1 units are replaced at 9 and 13 years, and the Type 2 units are replaced when their degradation level reaches the preventive maintenance threshold, 69 and 84, respectively.

At the beginning of the genetic algorithm, an initial population of chromosomes is generated as the ‘initial generation,’ and the initial solutions in this phase are generated randomly. In general, there are three genetic operators: crossover, mutation, and selection, which maintain balance between exploration and exploitation in the search space. In the crossover operation, two new chromosomes (offspring) are produced by combining two chromosomes (parents) that have survived from the previous generation, and a single-point crossover is used to generate new chromosomes as shown in Fig. 7.

In the mutation operation, the deterministic adaptation is considered and the mutation rate of the current generation will decrease from 0.5 to 0.2 as the number of generations increases [26]. The value of the genes in the chromosome is replaced depending on whether the target interval availability is satisfied or not. After performing the mutation operation, the heuristic method is used to improve the

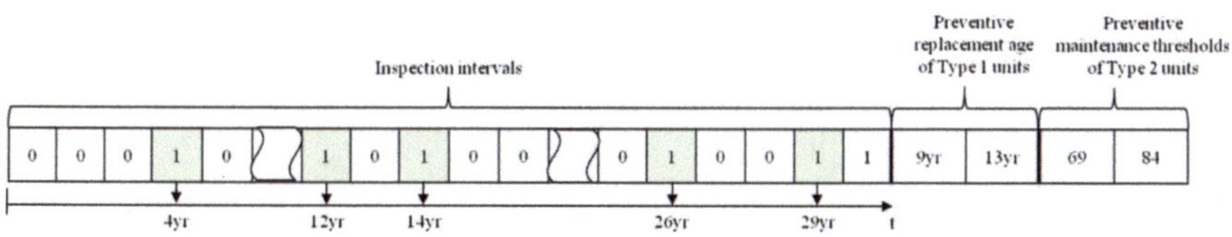

Fig. 6 An example of the solution representation in Phase 1

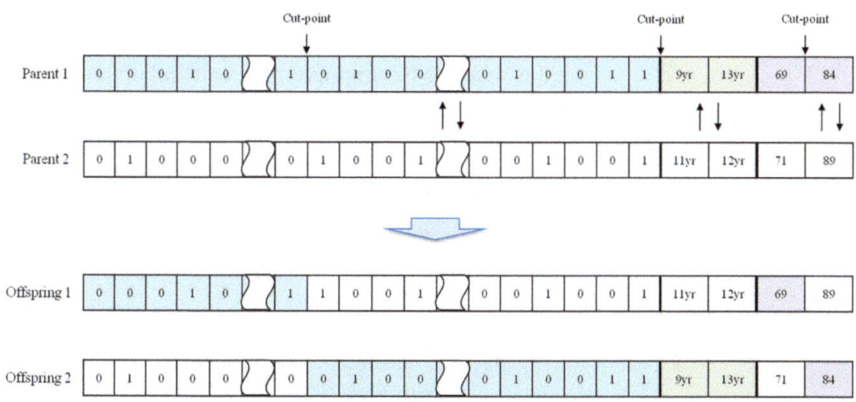

Fig. 7 An example of one-cut point crossover in Phase 1

alternatives in Phase 1. In the heuristic method, the target standard deviation of the interval availability over all periods is considered, and the heuristic procedure is terminated either if the target standard deviation is satisfied or if the number of replications for improving the alternative is reached. The detailed procedure is as follows:

Step 1 Select a chromosome which is not yet chosen and obtain the standard deviation of the interval availability by simulation.

1.1 If $A_{STD} \leq A_{STD\_T}$, terminate the procedure. Otherwise (if $A_{STD} > A_{STD\_T}$), generate the number of replications for improving the alternative and go to Step 2.

Step 2 Select the first period that does not satisfy the target interval availability and let the period be pth period.

2.1 If $AI_p < AI_T - \varepsilon$, improve the standard deviation of the interval availability as follows. Otherwise, go to Step 2.2.

2.1.1 Shorten the inspection interval of pth period as much as the unit time and obtain $\Delta A_{STD}/\Delta Cost$ by simulation.
2.1.2 Decrease the preventive replacement age of each Type 1 unit as much as the improvement unit and obtain $\Delta A_{STD}/\Delta Cost$ by simulation.
2.1.3 Decrease the preventive maintenance threshold of each Type 2 unit as much as the improvement unit and obtain $\Delta A_{STD}/\Delta Cost$ by simulation.
2.1.4 Improve the current standard deviation by the way with the highest score and go to Step 3.

2.2 If $AI_p > AI_T + \varepsilon$, improve the standard deviation of the interval availability as follows.

2.2.1 Lengthen the inspection interval of pth period as much as the unit time and obtain $\Delta A_{STD}/\Delta Cost$ by simulation.
2.2.2 Increase the preventive replacement age of each Type 1 unit as much as the improvement unit and obtain $\Delta A_{STD}/\Delta Cost$ by simulation.
2.2.3 Increase the preventive maintenance threshold of each Type 2 unit as much as the improvement unit and obtain $\Delta A_{STD}/\Delta Cost$ by simulation.
2.2.4 Improve the current standard deviation by the way with the lowest score and go to Step 3.

Step 3 Check where the target standard deviation is satisfied or not.

3.1 If $A_{STD} \leq A_{STD\_T}$, terminate the procedure. Otherwise (if $A_{STD} > A_{STD\_T}$), go to Step 3.2.

3.2 If the current replications are less than the number of replications generated in Step 1.1, increase the number of the current replications by one and go to Step 2. Otherwise, terminate the procedure.

3.3 Hybrid Genetic Algorithm with Heuristic Method in Phase 2

In Phase 2, the inspection schedule of one-shot systems with limited maintenance resources is considered, and the starting point of inspection of each one-shot system is determined in order to minimize the life cycle cost and satisfy the target interval availability. In this phase, 0.5 year is considered as the unit time and the chromosomes for the starting points are represented by integer strings and real numbers. Figure 8 shows an example of a solution representation when applied to 10 one-shot systems. For example, the optimal inspection time points for the first and second periods are assumed to be 4 and 7 years, respectively, which are determined in Phase 1, and the value of the first gene in the chromosome for starting points is 1.5 years. This means that the first and second inspection of the first one-shot system will be performed at 2.5 (= 4.0–1.5) years and 5.5 (= 7.0–1.5) years, respectively.

A single-point crossover is also used to generate a pair of offspring by exchanging the genes of parent chromosomes, as shown in Fig. 9.

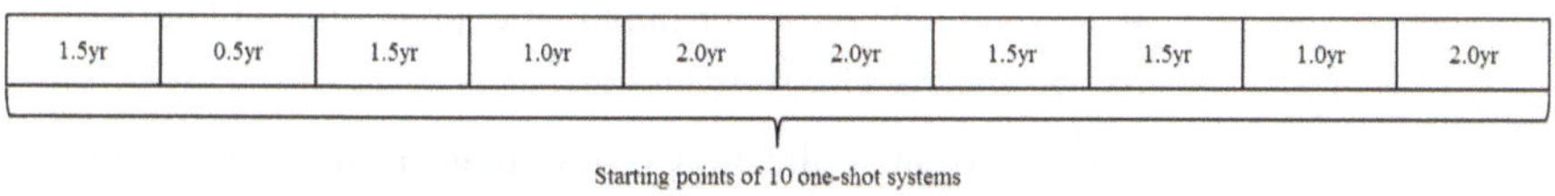

Fig. 8 An example of the solution representation in Phase 2

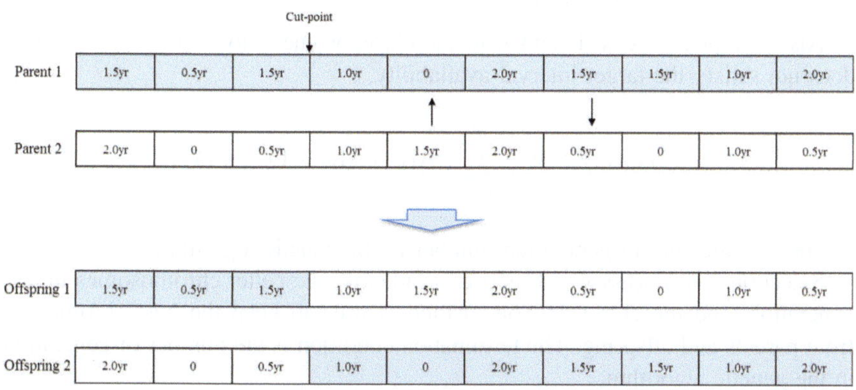

Fig. 9 An example of one-cut point crossover in Phase 2

In the mutation operation, the uniform random value, Θ, is used for each gene in the chromosome. If $\Theta \leq$ Mu, we randomly assign a new starting point to the value of the gene and the deterministic adaptation is also applied in Phase 2. After performing the mutation operation in Phase 2, a heuristic method is proposed to improve alternatives for satisfying the target standard deviation of the interval availability, and the detailed procedure is as follows:

Step 1 Select a chromosome which is not yet chosen.

Step 2 Estimate the standard deviation of the interval availability for each system by simulation.

- 2.1 If $A_{\text{STD}} \leq A_{\text{STD\_T}}$, terminate the procedure. Otherwise (if $A_{\text{STD}} > A_{\text{STD\_T}}$), generate the number of replications for improving the alternative and go to Step 3.

Step 3 Select a system which has the largest standard deviation of the interval availability among the systems.

- 3.1 Assign a new starting point which is not yet selected.
- 3.2 Obtain the standard deviation increment per unit cost $\Delta A_{\text{STD}}/\Delta\text{Cost}$, by simulation.
- 3.3 If all starting points are checked, replace the current starting point to the starting point with the lowest score and go Step 4. Otherwise, go to Step 3.1.

Step 4 Obtain the standard deviation of the interval availability by simulation.

- 4.1 If $A_{\text{STD}} \leq A_{\text{STD\_T}}$, terminate the procedure. Otherwise (if $A_{\text{STD}} > A_{\text{STD\_T}}$), go to Step 4.2.
- 4.2 If the current replication is less than the number of replications generated in Step 3.2, increase the number of the current replications by one and go to Step 3. Otherwise, terminate the procedure.

Also, the penalty cost is added to the chromosome whose interval availability does not satisfy the target interval availability.

$$\text{Penalty cost} = \sum_{p=1}^{P} (|\text{AI}_{\text{T}} - \text{AI}_{P}| \times \text{LC}) \times \omega \tag{2}$$

where ω is the current generation number in the genetic algorithm.

Selection is a process which is used to identify the better chromosomes in each generation, and the elite selection method is used to keep the best chromosome from parents and offspring. The termination condition is the number of generations in the genetic algorithm.

3.4 Heuristic Method in Phase 2

In Phase 2, a heuristic method is also proposed to compare the performance of the proposed hybrid genetic algorithm. The detailed procedure of the heuristic method is as follows:

Step 1 Set the starting point of all one-shot systems as '0.'
Step 2 Estimate the interval availability and life cycle cost by simulation.
Step 3 Check whether the target interval availability is satisfied or not.

3.1 If $AI_P \geq AI_T$ (for all p), select the current alternative and terminate the procedure.

3.2 If $AI_P < AI_T$ for at least one period, set the current starting point as '0' and go to Step 4.

Step 4 Improve the interval availability through putting the earliest first inspection point forward.

4.1 If $IP_{Earliest} = 0.5$, go to Step 5. Otherwise, go to Step 4.2.

4.2 Generate a new starting point, $ST_{Alt\_Forward}$ as follows and add it to the current starting points.

$$ST_{Alt\_Forward} = (IP_{Earliest} - IP_{Earliest}) + 0.5$$

4.3 Assign the current starting points to the systems equally.

4.4 Obtain the increment availability per unit cost by simulation.

4.5 Delete the starting point generated in Step 4.1 from the current starting points and go to Step 5.

Step 5 Improve the interval availability through putting the latest first inspection point backward.

5.1 If $IP_{Latest} = IP_{Opt\_Second} - 0.5$, go to Step 6. Otherwise, go to Step 5.2.

5.2 Generate a new starting point, $ST_{Alt\_Backward}$ as follows and add it to the current starting points.

$$ST_{Alt\_Backward} = (IP_{Latest} - IP_{Opt\_First}) + 0.5$$

5.3 Assign the current starting points to the systems equally.

5.4 Obtain the increment availability per unit cost by simulation.

5.5 Delete the starting point generated in Step 5.1 from the current starting points and go to Step 6.

Step 6 Select the alternative with the highest score.

6.1 If both alternatives generated in Step 4.1 and 5.1 satisfy the target interval availability, select the alternative with the highest score and terminate the procedure. Otherwise, go to Step 6.2.

6.2 If one of the alternatives satisfies the target interval availability, select the alternative and terminate the procedure. Otherwise, go to Step 6.3.

6.3 Select the alternative with the highest score.

6.3.1 Add the starting point to the current starting point candidates.

6.3.2 Change IP_{First} and IP_{Latest} based on the selected alternative and go to Step 4.

4 Numerical Examples

For numerical examples, we consider 10 one-shot systems consisting of 10 units (A to J) with a series structure and the system lifetime is 30 years (262,800 h). Table 3 shows the failure and repair distributions, CM and PM costs of Type 1 units, and we assume that the time to failure of Type 1 units follows either an exponential distribution or a Weibull distribution. In the system, Units *I* and *J* are Type 2 units,

Table 3 Failure and repair distributions, CM and PM costs of units

Name	Failure distribution (year)	Repair distribution (h)	CM cost	PM cost
A	Exp (27.5)	Exp (12)	350,000	–
B	Exp (26.0)	Exp (12)	320,000	–
C	Exp (28.0)	Exp (10)	270,000	–
D	Exp (25.5)	Exp (10)	300,000	–
E	Exp (29.0)	Exp (4)	280,000	–
F	Exp (25.0)	Exp (6)	340,000	–
G	Weibull (27.0, 1.8)	Exp (6)	280,000	190,000
H	Weibull (26.5, 2.0)	Exp (4)	330,000	170,000
I	–	Exp (10)	360,000	200,000
J	–	Exp (12)	290,000	210,000

and the gamma process is considered to describe their degradation process. The shape parameter of the gamma distribution is $\alpha = c \cdot t^b$, where $c = 1.1$ and $b = 0.05$ for Unit *I*, while $c = 1.2$ and $b = 0.01$ for Unit *J*. The scale parameter of the gamma distribution, β, are 0.001 and 0.002 for Units *I* and *J*, and the failure thresholds are 110 and 130, respectively. The set-up cost per inspection is 700,000, and the inspection operation cost per system is 1,000,000. The total number of maintenance equipment is five, and a maintenance equipment is needed to perform the maintenance of one unit. The simulation length is 30 years (262,800 h), and the number of replications is 10. The parameters for the genetic algorithm in Phase 1 are as follows: crossover rate (0.7), mutation rate (0.5), population size (50), and generation size (50). The parameters for the genetic algorithm in Phase 2 are as follows: crossover rate (0.7), mutation rate (0.5), population size (50), and generation size (70). For the heuristic method, the improvement unit is 10%, the allowable gap, ε, is 0.01 and the target standard deviation of the inspection interval availability is 0.02.

For numerical examples, we firstly consider different target interval availability, 0.75, 0.80, and 0.85 (***CASE 1***), and decrease the scale parameter of the failure distribution of the Type 1 units by 15, 30 and 45% (***CASE 2***). Lastly, the failure threshold of the Type 2 units is increased by 15, 30, and 45% (***CASE 3***). From numerical results, we analyze the effect of model parameters on the optimal solutions and compare the general genetic algorithm with the hybrid genetic algorithm and the heuristic method. To evaluate the performance of the general genetic algorithm, hybrid genetic algorithm, and heuristic method, they are coded in C++ programming language and the numerical examples are executed on an IBM-PC compatible with an Intel Core 3.3 GHz.

4.1 CASE 1: Different Target Interval Availability

Firstly, we consider different target interval availability, 0.75, 0.80, and 0.85, and Fig. 10 shows that the life cycle cost increases as the target interval availability increases because the inspection of one-shot systems should be carried out more frequently in order to satisfy the high target interval availability. The hybrid genetic algorithm results in lower life cycle cost than both the general genetic algorithm and the heuristic method.

Table 4 shows the optimal inspection points of a one-shot system in Phase 1. The total number of inspections in Phase 1 increases as the target interval availability increases because one-shot systems must be inspected more frequently in order to reduce the time interval between system failure and detection. In order to satisfy the high target interval availability, it is also necessary to improve the reliability of the units through more frequent preventive maintenance, thus the preventive replacement age of Units *G* and *H*, and preventive maintenance threshold of Units *I* and *J* should decrease, respectively.

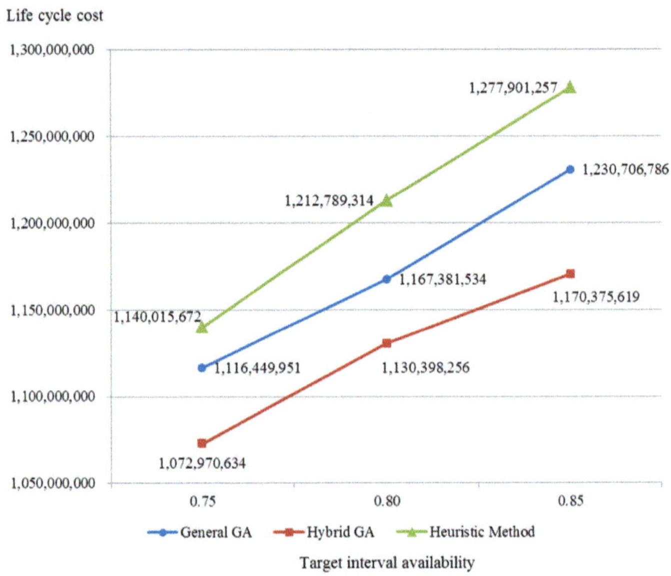

Fig. 10 Life cycle cost for different target interval availability

Table 4 Optimal inspection points, preventive replacement ages, and preventive maintenance thresholds in Phase 1

Target interval availability	Inspection points (year)	Preventive replacement age (year)		Preventive maintenance threshold	
		G	*H*	*I*	*J*
0.75	4, 7, 11, 23, 24	25.96	23.33	92	73
0.80	4, 7, 10, 19, 25, 27	23.54	21.83	89	63
0.85	4, 5, 13, 19, 20, 22, 23, 28, 29	21.94	20.07	72	57

Table 5 shows the optimal starting points of inspection of one-shot systems and CPU times to obtain the solutions in Phase 2. The total number of inspections in Phase 2 increases as the target interval availability increases because frequent inspection is needed in order to keep the reliability of one-shot systems in a high level.

Table 5 Optimal starting points of one-shot systems and CPU times in Phase 2

Target interval availability	Method	Starting point (the number of the system) (year)	Total inspection numbers	CPU time (s)
0.75	Heuristic	0.0(2), 0.5(2), 1.0(2), 1.5(2), 2.0(1), −0.5(1)	26	4237
	General GA	0.0 (2), 1.0(1), 1.5(2), 2.0 (2), −0.5(3)	23	8067
	Hybrid GA	0.0 (1), 1.0(1), 2.0(2), 2.5(3), −2.5(3)	21	14,895
0.80	Heuristic	0.0(2), 0.5(2), 1.0(2), 1.5(2), −0.5(1), −1.0(1)	34	4492
	General GA	0.0(1), 0.5(2), 1.5(2), −0.5(3), −1.0(2)	29	8481
	Hybrid GA	0.5(1), 1.5(2), 2.0(3), 3.5(2), −1.0(2)	25	15,067
0.85	Heuristic	0.0(2), 0.5(2), 1.0(2), −0.5(2), −1.0(1), −1.5(1)	36	5082
	General GA	0.0(1), 1.0(2), 1.5(2), 2.0(3), −1.0(2)	31	8867
	Hybrid GA	1.0(3), 2.0(2), 2.5(2), 3.5(3)	28	15,607

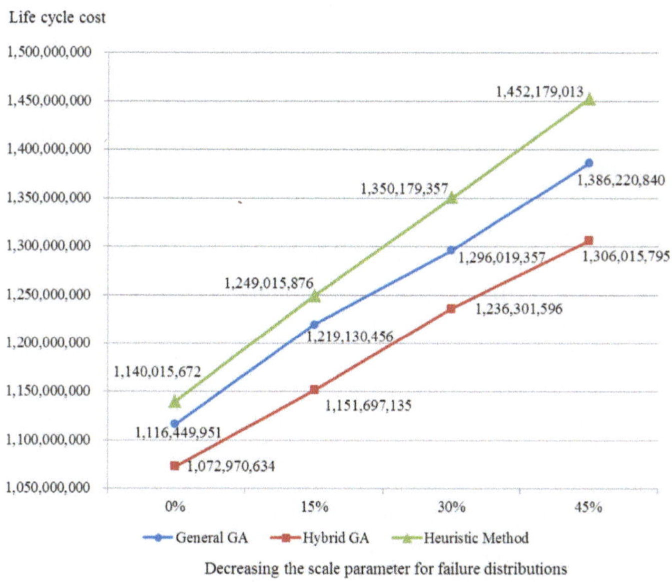

Fig. 11 Life cycle cost for different scale parameters of failure distribution of Type 1 units

Table 6 Optimal inspection points, preventive replacement ages, and preventive maintenance thresholds in Phase 1

Decrement (%)	Inspection points (year)	Preventive replacement age (year)		Preventive maintenance threshold	
		G	*H*	*I*	*J*
0	4, 7, 11, 23, 24	25.96	23.33	92	73
15	4, 7, 10, 19, 25, 27	23.54	21.83	90	71
30	4, 5, 13, 19, 20, 22, 23, 28, 29	18.11	18.09	95	74
45	3, 6, 13, 17, 19, 20, 22, 23, 24, 26, 27	17.39	17.52	93	69

4.2 CASE 2: Decreasing the Scale Parameters of the Failure Distribution of Type 1 Units

In this case, we decrease the scale parameter of failure distribution of Type 1 units by 15, 30, and 45%. Figure 11 shows that the life cycle cost increases as the scale parameter of the failure distribution decreases because the failure of Type 1 units occurs more frequently and the time interval between the system failure and detection by inspection may be increased. Thus, the time interval should be reduced through frequent inspection with high maintenance costs in order to satisfy the high target interval availability. Also, the hybrid genetic algorithm results in a lower life cycle cost than either the general genetic algorithm or the heuristic method.

Table 6 shows the optimal inspection points of a one-shot system in Phase 1. Because of decreasing reliability of Type 1 units, one-shot systems should be inspected more frequently during their lifetime to detect the system failure quickly. As a result, the total number of inspection of one-shot systems increases as the scale parameter of failure distribution of Type 1 units decreases. Also, the increased failure of Type 1 units may reduce *MUT* (mean up time) between inspection times, therefore, the reliability of Units *G* and *H* should be improved through more frequent preventive maintenance. Hence, the preventive replacement ages of Units *G* and *H* increases as the scale parameters of the failure distribution of Units *G* and *H* decreases. However, the preventive maintenance thresholds of Units *I* and *J* do not change much.

Table 7 shows the optimal starting points of one-shot systems and CPU times required to obtain the solutions. As the reliability of Type 1 units decreases, more frequent inspection of one-shot systems is needed to detect the system failure quickly. As a result, the total number of inspection of one-shot systems increases as the scale parameter of failure distribution of Type 1 units decreases.

Table 7 Optimal starting points of one-shot systems and CPU times in Phase 2

Decrement (%)	Method	Starting point (the number of the system) (year)	Total inspection numbers	CPU time (s)
0	Heuristic	0.0(2), 0.5(2), 1.0(2), 1.5(2), 2.0(1), −0.5(1)	26	4237
	General GA	0.0 (2), 1.0(1), 1.5(2), 2.0 (2), −0.5(3)	23	8067
	Hybrid GA	0.0 (1), 1.0(1), 2.0(2), 2.5(3), −2.5(3)	21	14,895
15	Heuristic	0.0(2), 0.5(2), 1.0(2), 1.5(2), −0.5(2)	29	5097
	General GA	0.0(2), 0.5(2), 1.0(1), 2.5(3), −0.5(2)	27	8517
	Hybrid GA	0.0(1), 1.0(2), 1.5(2), 3(2), −2.0 (3)	24	15,196
30	Heuristic	0.0(2), 0.5(2), 1.0(2), 1.5(2), −0.5(1), −1.0(1)	36	5437
	General GA	0.5(2), 1.5(1), 2.0(3), −0.5(2), −1.5(2)	31	8974
	Hybrid GA	0.5(2), 1.5(3), 2.0(4), −1.5(3)	30	15,387
45	Heuristic	0.0(2), 0.5(2), 1.0(2), −0.5(2), −1.0(1), −1.5(1)	44	5542
	General GA	0.5(2), 1.5(2), 3.0(2), −1.0(2), −1.5(2)	39	10,837
	Hybrid GA	0.5(2), 1.0(2), 2.0(2), −1.0(2), −2.0(2)	35	15,731

4.3 CASE 3: Increasing the Failure Thresholds of Type 2 Units

If we increase the failure threshold of Type 2 units by 15, 30, and 45%, the life cycle cost decreases as shown in Fig. 12 because the time to failure of Type 2 units becomes long and hence, they do not require such frequent inspection. Also, the hybrid genetic algorithm results in lower life cycle cost than the general genetic algorithm or the heuristic method.

Table 8 shows the optimal inspection points of a one-shot system when the failure thresholds of Units *I* and *J* increase. Increasing the failure thresholds of Units *I* and *J* means that the time to failure of Units I and J increases and it results that the *MUT* between inspection times increases. Hence, the interval availability can satisfy the target availability even though the inspection of one-shot systems is performed less frequently, and the total number of inspection decreases as the failure thresholds of Units *I* and *J* increases. The preventive maintenance of Unit *G* and *H* does not need to be performed so frequently. However, it is necessary to

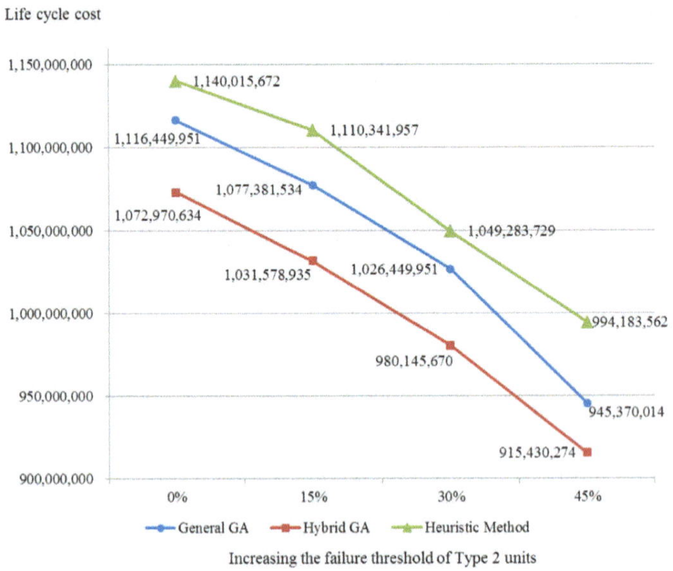

Fig. 12 Life cycle cost for different failure thresholds of Type 2 units

Table 8 Optimal inspection points, preventive replacement ages, and preventive maintenance thresholds in Phase 1

Increment (%)	Inspection points (year)	Preventive replacement age (year)		Preventive maintenance threshold	
		G	*H*	*I*	*J*
0	4,7,11,23,24	25.96	23.33	92	73
15	4,10,16,25,27	25.89	23.21	103	91
30	4,14,16,20	26.06	23.31	111	116
45	4,13,20,27	25.91	23.18	124	126

carry out the preventive maintenance of Units *I* and *J* more frequently in order to improve their reliabilities.

Table 9 shows the optimal starting points of inspection of one-shot systems and CPU times when the failure thresholds of Units *I* and *J* increase. As the failure of thresholds of Units *I* and *J* increase, the time to failures of Units *I* and *J* increases, and it results that the *MUT* between inspection times increases. Thus, the target interval availability can be satisfied even though the systems are inspected less frequently, and the total number of inspection decreases as the failure threshold of Units *I* and *J* increases.

Table 9 Optimal starting points of one-shot systems and CPU times in Phase 2

Increment (%)	Method	Starting point (the number of the system) (year)	Total inspection numbers	CPU time (s)
0	Heuristic	0.0(2), 0.5(2), 1.0(2), 1.5(2), 2.0 (1), −0.5(1)	26	4237
	General GA	0.0 (2), 1.0(1), 1.5(2), 2.0 (2), −0.5(3)	23	8067
	Hybrid GA	0.0 (1), 1.0(1), 2.0(2), 2.5(3), −2.5 (3)	21	14,895
15	Heuristic	0.0(2), 0.5(2), 1.0(2), −0.5(2), −1.0(2)	24	4018
	General GA	0.5(2), 1.0(1), 2.5(2), −1.0(3), −1.5(2)	22	7832
	Hybrid GA	0.5(2), 1.0(3), 2.5(3), −1.5(2)	18	12,905
30	Heuristic	0.0(2), 0.5(2), 1.0(2), −0.5(2), −1.0(2)	19	3709
	General GA	1.0(2), 1.5(2), 2.5(3), −2.0(3)	16	7537
	Hybrid GA	1.0(2), 1.5(2), −0.5(3), −1(3)	14	12,137
45	Heuristic	0.0(3), 0.5(3), −0.5(2), −1.0(2)	16	3569
	General GA	1.0(3), 2.0(4), −0.5(3)	12	7132
	Hybrid GA	1.0(3), 2.5(3), −1.5(4)	11	11,518

5 Conclusions

In this chapter, we considered an inspection scheduling problem of one-shot systems. In practice, numerous one-shot systems are stored in operational sites until their inspection or operation. In order to perform the inspection of one-shot systems, they are moved to the maintenance sites, each of which contains a certain amount of maintenance resources, such as spare parts, engineers, and maintenance equipment. Therefore, the maintenance delay time due to a shortage of maintenance resources may increase if all one-shot systems are inspected and maintained at the same time, and thus, we need to establish the efficient inspection schedule of one-shot systems. In this chapter, we dealt with an inspection schedule problem for one-shot systems under a constraint of maintenance resources. The one-shot system usually consists of two types of units, where Type 1 units fail at random times, and Type 2 units are degraded with time. For the evaluation of one-shot system performance quantitatively, the interval availability between inspection points was used as an optimization criterion, because they perform a mission suddenly between inspections and once only during the life cycle. Also, the inspection and maintenance of one-shot systems incurs maintenance costs, and then the life cycle cost was considered as an optimization criterion. Simulation was used to estimate the interval availability and life cycle cost and a hybrid genetic algorithm with a heuristic method for the both phases, and a heuristic method for Phase 2 were proposed in order to find near-optimal inspection schedule of one-shot systems.

Also, the inspection intervals and preventive replacement age of Type 1 units, as well as the preventive maintenance threshold of Type 2 units and the starting point of inspection of one-shot systems, were determined to minimize the life cycle cost and satisfy the target interval availability through two phases. In numerical examples, the results showed that the inspection of one-shot systems is performed frequently and the preventive replacement age of Type 1 units and preventive maintenance threshold of Type 2 units decrease as the target interval availability increases and the scale parameter of failure distribution of Type 1 units increase. However when the failure threshold of Type 2 units increases, the total number of inspections decreases and preventive maintenance of Type 2 units are performed infrequently because the time to failure of Type 2 units increases. Also, the hybrid genetic algorithm with a heuristic method gives better alternatives than the general genetic algorithm or the heuristic method.

For further studies, we can consider real-sized problems numerically and propose more efficient meta-heuristics.

Acknowledgements This research was supported by the Basic Science Research Program through the National Research Foundation of Korea (NRF), funded by the Ministry of Education, Science and Technology (NRF-2016R1D1A1A09919337).

References

1. Nakagawa T, Mizutani S (2009) A summary of maintenance policies for a finite interval. Reliab Eng Syst Safety 94:89–96
2. Nakagawa T, Mizutani S, Chen M (2010) A summary of periodic and random inspection policies. Reliab Eng Syst Safety 95:906–911
3. Ito K, Nakagawa T (1992) Optimal inspection policies for a system in storage. Comput Math Appl 24:87–90
4. Kaio N, Dohi T, Osaki S (1994) Inspection policy with failure due to inspection. Microelectron Reliab 34:599–602
5. Ito K, Nakagawa T, Nishi K (1995) Extended optimal inspection policies for a system in storage. Math Comput Model 22:83–87
6. Ito K, Nakagawa T (1995) An optimal inspection policy for a storage system with high reliability. Microelectron Reliab 35:875–882
7. Ito K, Nakagawa T (1995) An optimal inspection policy for a storage system with three types of hazard rate functions. J Oper Res Soc Jpn 38:423–431
8. Ito K, Nakagawa T (2000) Optimal inspection policies for a storage system with degradation at periodic tests. Math Comput Model 31:191–195
9. Taghipour S, Banjevic D (2011) Periodic inspection optimization models for a repairable system subject to hidden failures. IEEE Trans Reliab 60:275–285
10. Golmakani HR, Moakedi H (2012) Optimal non-periodic inspection scheme for a multi-component repairable system using A* search algorithm. Comput Ind Eng 63:1038–1047
11. Leung FK (2001) Inspection Schedules when the lifetime distribution of a single-unit system is completely unknown. Eur J Oper Res 132:106–115
12. Cui L, Xie M (2005) Availability of a periodically inspected system with random repair or replacement times. J Stat Plan Inference 131:89–100

13. Kitagawa T, Yuge T, Yanagi S (2015) Optimal maintenance policy of a multi-unit one-shot system with minimal repair. IEICE Trans Fundam 98-A:2077–2083
14. Kitagawa T, Yuge T, Yanagi S (2016) Periodic and non-periodic inspection policies for a one-shot system with minimal repair. J Jpn Ind Manag Assoc 66:385–387
15. Yun WY, Han YJ, Kim HW (2012) Simulation-based inspection policies for a one-shot system. In: Proceedings of the 5th Asia-Pacific international symposium, Nanjing, China, pp. 621–628
16. Grall A, Berenguer C, Dieulle L (2002) A condition-based maintenance policy for stochastically deteriorating systems. Reliab Eng Syst Safety 76:167–180
17. Avramidis AN, L'ecuyer P, Tremblay P-A (2003) Efficient simulation of gamma and variance-gamma processes. In: Proceedings of the 2003 winter simulation conference, 1, pp. 319–326
18. Li W, Pham H (2005) An inspection-maintenance model for systems with multiple competing processes. IEEE Trans Reliab 54:318–327
19. Van Noortwijk JM (2009) A survey of the application of gamma processes in maintenance. Reliab Eng Syst Safety 94:2–21
20. Yun WY, Han YJ, Kim HW (2013) Optimal inspection policies for a one-shot system with two types of units. In: Proceedings of the 9th international conference on intelligent manufacturing and logistics systems, Shanghai, China, pp. 136–141
21. Yun WY, Liu L, Han YJ (2014) Metaheuristic-based inspection policy for a one-shot system with two types of units. J Mech Sci Technol 28:3947–3955
22. Yun WY, Liu L, Han YJ, Rhee DW, Han CG (2013) Optimal inspection schedules for one-shot systems with two types of units. In: Proceedings of the international conference on quality, reliability, risk, maintenance and safety engineering, pp. 631–635
23. Yun WY, Liu L, Han YJ, Rhee DW, Han CG (2013) Simulation-based optimal inspection schedules for one-shot systems with two types of units. In: Proceedings of the 17th international conference on industrial engineering: theory, application and practice, pp. 314–320
24. Yun WY, Moon IK, Kim KR (2008) Simulation-based maintenance support system for multi-functional complex systems. Prod Plan Control 19:365–378
25. Gen M, Cheng R, Lin L (2008) Network models and optimization: multi-objective genetic algorithm approach. Springer, London
26. Yun WY, Han YJ, Kim HW (2014) Simulation-based inspection policies for a one-shot system in storage over a finite time span. Commun Stat Simulat Comput 43:1979–2003

Industrial System Performance Under Multistate Failures with Standby Mode

Amit Kumar, Mangey Ram, Sangeeta Pant and Anuj Kumar

1 Introduction

In many real-life situations, system reliability plays very important role. To get the settled goal of production, one has to keep the system free from all possible failures. High reliability of the system increases the efficiency of system performance [1, 2]. Also, it is well known that the reliability measures are of foremost concern in the planning, design, and operation of any system or equipment. Many researchers have done an elite research in the field of system reliability [3–7]. In the present age, as competition increases, complexity of industrial system also increased. So in order to compete to the other similar industry, the system must be free from failures as much as possible. Standby technique is one of the techniques which can overcome this issue [8, 9]. Misra and Verma [10] found that reliability and risk assessment are the most efficient, effective tools that assists in decision making for safety and risk management of complex engineering systems, such as in aerospace, nuclear power plants, and chemical process industries. Dhillon [11] presented reliability and availability analysis of a two-unit parallel system with warm standby and common cause failure. The failed unit's repair time is assumed to be arbitrary distributed.

Pan [12] discussed a multistage production system with one machine/tool in active state and n spares in standby state. He found that when the operating machine/tool breaks down, a switching device detects the machine failure via the sensor and the defective tool is replaced with a functional spare, so that the system

A. Kumar (✉)
Department of Mathematics, Lovely Professional University, Punjab, India
e-mail: amit303singh@gmail.com

M. Ram
Department of Mathematics, Graphic Era University Dehradun, Uttrakhand, India

S. Pant · A. Kumar
Department of Mathematics, University of Petroleum & Energy Studies, Dehradun, India

M. Ram and J.P. Davim (eds.), *Modeling and Simulation in Industrial Engineering*, Management and Industrial Engineering, DOI 10.1007/978-3-319-60432-9_4

can resume its operation. Azaron et al. [13] discussed a L-dissimilar-unit non-repairable cold standby redundant system, in which each unit is composed of a number of independent components with generalized Erlang distribution of lifetimes, arranged in any general configuration.

Filieri et al. [14] focused on a computer-based software system that operate in safety-critical environment, where a relevant quality factor, system reliability, is defined as a probabilistic measure of the system's ability to successfully carry out its own task. El-Said and El-Sherbeny [15] investigated the cost-benefit analysis of a two-unit cold standby system with two-stage repair of a failed unit and discussed the concept of waiting time with two-stage repair for a two-unit cold standby system. Researchers including [16–18] analyzed and evaluated the reliability measures for various engineering models under the concept of Gumbel-Hougaard family copula with different repair policies. Here the authors analyzed the system performance in terms of various reliability measures. Gupta and Tewari [19] discussed the development of a power generation system of a thermal plant by usingMarkov technique with probabilistic approach. This study covered two areas: development of a predictive model and evaluation of performance with the help of developed model. Majeed and Sadiq [20] investigated a Dokon hydropower station by using Markov technique. An enormous research has been done in the past regarding human error, complex systems [21–25] as human error plays a crucial role in operating any industrial systems.

System reliability occupies progressively more significant use in industry, manufacturing, design, etc. Maintaining a high or required level of accuracy is often an essential requirement of a system [26, 27]. The study of the repairmen is an essential part of a repairable system and can affect the economy of the system directly or indirectly. Therefore, his/her action and work frames are vital in improving the reliability of a repairable system. Keeping all the above researches and facts in mind, here the authors try to investigate a complex industrial system which consist of n units with a human operator and find various factors which affects the system performance.

2 Problem Statement

In the present chapter, a complex industrial system is considered which consists of n units connected in parallel configuration with a standby unit operated by a human operator. The designed type industrial system can be easily found in a chapter plant system, sugar industry-based system, hydro power plant system, and many other engineering fields. Four types of failures have exists in the considered system throughout the task, namely unit failure, catastrophic failure, human failure, and standby unit failure. When all the parallel units of the system have failed, then the standby unit will start working automatically. The system goes to complete failed state due to human error, catastrophic failure, all parallel unit failure, and the failure of standby unit. The system can repair after a unit failure (partial failure) as well as

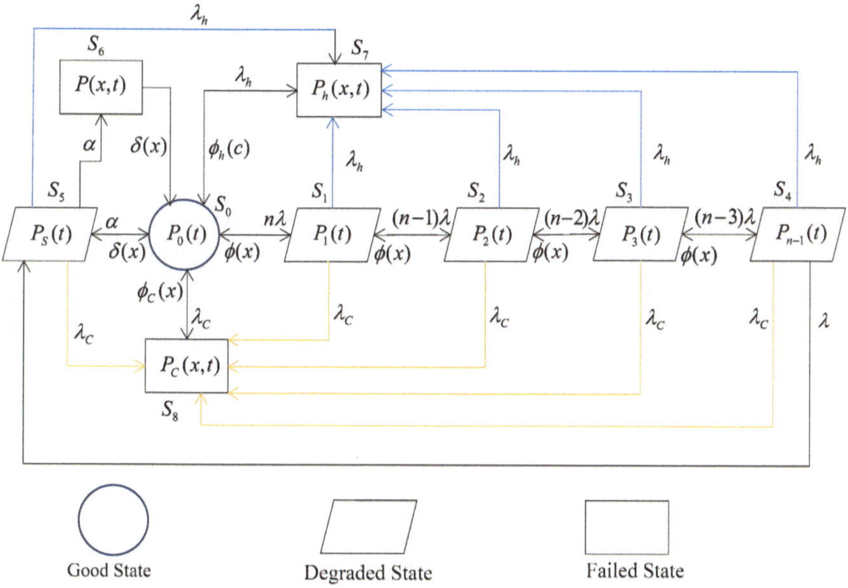

Fig. 1 Transition state diagram

the major failure (complete failure). After repair, the system works like a new one. The system may work with reduced capacity i.e. it works in degraded state also. The failure and repair rates of the system are taken to be constant.

The transition state diagram of the proposed model has been shown in Fig. 1.

Where, $P_0(t)$, $P_1(t)$, $P_2(t)$, $P_3(t)$, $P_{n-1}(t)$, $P_s(t)$ are the probabilities that at time t, the system is working with full capacity, with one failed unit, with two failed units, with three failed units, with $(n-1)$ failed units, working with standby unit, respectively. $P_C(x, t)$, $P_h(x, t)$, $P(x, t)$ are the probabilities that the system is in a completely failed state due to catastrophic failure; human failure, the failure of standby unit; the system is running under repair with elapsed repair time x, t. λ, λ_C, λ_h, and α are the failure rate of parallel units, catastrophic failure, human error, and standby unit failure rate, respectively. $\phi(x)$, $\phi_h(x)$, ϕ_C, $\delta(x)$ are the repair rates of parallel unit, human error, catastrophic failure, and the failure of standby unit, respectively.

State Description, Table 1.

3 Mathematical Formulation and Solution

At any time t, if the system is in state S_i, then the probability of the system to be in that state is defined as: Probability that the system is in state S_i at time t and remains there in interval $(t, t + \Delta t)$ or/and if it is in some other state at time t then it should

Table 1 State description

S_0	The system is in good state
S_1	The system is working in degraded state with one failed unit
S_2	The system is working in degraded state with two failed unit
S_3	The system is working in degraded state with three failed unit
S_4	The system is working in degraded state with $(n-1)$ failed unit
S_5	The system is working in degraded state with failure of standby unit
S_6	The system is failed due to failure of standby unit after failure of all of n parallel units
S_7	The system is failed due to human error
S_8	The system is failed due to catastrophic failure

transit to the state S_i in the interval $(t, t+\Delta t)$ provided transition exist between the states as $\Delta t \to 0$. Accordingly, the following set of intro-differential equations is formed as:

The probability of the system to be in state S_0 in the interval $(t, t+\Delta t)$ is given by

$$P_0(t+\Delta t) = (1 - n\lambda\Delta t)(1 - \lambda_h\Delta t)(1 - \lambda_C\Delta t)(1 - \alpha\Delta t)P_0(t) + \delta(x)P_S(t)\Delta t$$
$$+ \phi(x)P_1(t)\Delta t + \int_0^\infty \phi_h(x)P_h(x,t)\Delta t\mathrm{d}x + \int_0^\infty \delta(x)P(x,t)\Delta t\mathrm{d}x + \int_0^\infty \phi_C(x)P_C(x,t)\Delta t\mathrm{d}x$$

$$\Rightarrow \frac{P_0(t+\Delta t) - P_0(t)}{\Delta t} + (n\lambda + \lambda_h + \lambda_C + \alpha)P_0(t) = \delta(x)P_S(t) + \phi(x)P_1(t)$$
$$+ \int_0^\infty \phi_h(x)P_h(x,t)\mathrm{d}x + \int_0^\infty \delta(x)P(x,t)\mathrm{d}x + \int_0^\infty \phi_C(x)P_C(x,t)\mathrm{d}x$$

Now, taking $\lim\limits_{\Delta t \to 0}$, we get:

$$\Rightarrow \lim_{\Delta t \to 0} \frac{P_0(t+\Delta t) - P_0(t)}{\Delta t} + (n\lambda + \lambda_h + \lambda_C + \alpha)P_0(t) = \delta(x)P_S(t)$$
$$+ \phi(x)P_1(t) + \int_0^\infty \phi_h(x)P_h(x,t)\mathrm{d}x + \int_0^\infty \delta(x)P(x,t)\mathrm{d}x + \int_0^\infty \phi_C(x)P_C(x,t)\mathrm{d}x$$

$$\left[\frac{\partial}{\partial t} + n\lambda + \lambda_h + \lambda_C + \alpha\right]P_0(t) = \delta(x)P_s(t) + \phi(x)P_1(t)$$
$$+ \int_0^\infty \phi_h(x)P_h(x,t)\mathrm{d}x + \int_0^\infty \delta(x)P(x,t)\mathrm{d}x + \int_0^\infty \phi_C(x)P_C(x,t)\mathrm{d}x \quad (1)$$

For state S_1,

$$\begin{aligned}
P_1(t+\Delta t) &= (1-\phi(x)\Delta t)(1-\{n-1\}\lambda\Delta t)(1-\lambda_C\Delta t)(1-\lambda_h\Delta t)P_1(t) \\
&\quad + n\lambda_1\Delta t P_0(t) + \phi(x)\Delta t P_2(t) \\
\Rightarrow \frac{P_1(t+\Delta t)-P_1(t)}{\Delta t} &+ (\phi(x)+\{n-1\}\lambda+\lambda_C+\lambda_h)P_1(t) = n\lambda_1 P_0(t) + \phi(x)P_2(t)
\end{aligned}$$

Now, taking $\lim\limits_{\Delta t\to 0}$, we get:

$$\begin{aligned}
&\lim_{\Delta t\to 0}\frac{P_1(t+\Delta t)-P_1(t)}{\Delta t} + (\phi(x)+\{n-1\}\lambda+\lambda_C+\lambda_h)P_1(t) = n\lambda_1 P_0(t) + \phi(x)P_2(t) \\
&\left[\frac{\partial}{\partial t} + \phi(x) + (n-1)\lambda + \lambda_C + \lambda_h\right]P_1(t) = n\lambda P_0(t) + \phi(x)P_2(t)
\end{aligned} \tag{2}$$

For state S_2,

$$\begin{aligned}
P_2(t+\Delta t) &= (1-\phi(x)\Delta t)(1-\{n-2\}\lambda\Delta t)(1-\lambda_C\Delta t)(1-\lambda_h\Delta t)P_2(t) \\
&\quad + (n-1)\lambda\Delta t P_1(t) + \phi(x)\Delta t P_3(t) \\
\frac{P_2(t+\Delta t)-P_2(t)}{\Delta t} &+ (\phi(x)+\{n-2\}\lambda+\lambda_C+\lambda_h)P_2(t) = (n-1)\lambda P_1(t) + \phi(x)P_3(t)
\end{aligned}$$

Now, taking $\lim\limits_{\Delta t\to 0}$, we get:

$$\begin{aligned}
&\lim_{\Delta t\to 0}\frac{P_2(t+\Delta t)-P_2(t)}{\Delta t} + (\phi(x)+\{n-2\}\lambda+\lambda_C+\lambda_h)P_2(t) = (n-1)\lambda P_1(t) + \phi(x)P_3(t) \\
&\left[\frac{\partial}{\partial t} + \phi(x) + (n-2)\lambda + \lambda_C + \lambda_h\right]P_2(t) = (n-1)\lambda P_1(t) + \phi(x)P_3(t)
\end{aligned} \tag{3}$$

For state S_3,

$$\begin{aligned}
P_3(t+\Delta t) &= (1-\phi(x)\Delta t)(1-\{n-3\}\lambda\Delta t)(1-\lambda_C\Delta t)(1-\lambda_h\Delta t)P_3(t) \\
&\quad + (n-2)\lambda\Delta t P_2(t) + \phi(x)\Delta t P_{n-1}(t) \\
\Rightarrow \frac{P_3(t+\Delta t)-P_3(t)}{\Delta t} &+ (\phi(x)+\{n-3\}\lambda+\lambda_C+\lambda_h)P_3(t) = (n-2)\lambda P_2(t) + \phi(x)P_{n-1}(t)
\end{aligned}$$

Now, taking $\lim\limits_{\Delta t \to 0}$, we get:

$$\lim_{\Delta t \to 0} \frac{P_3(t+\Delta t) - P_3(t)}{\Delta t} + (\phi(x) + \{n-3\}\lambda + \lambda_C + \lambda_h)P_3(t) = (n-2)\lambda P_2(t) + \phi(x)P_{n-1}(t)$$
$$\left[\frac{\partial}{\partial t} + \phi(x) + (n-3)\lambda + \lambda_C + \lambda_h\right]P_3(t) = (n-2)\lambda P_2(t) + \phi(x)P_{n-1}(t) \tag{4}$$

For state S_4,

$$P_{n-1}(t+\Delta t) = (1-\phi(x)\Delta t)(1-\lambda\Delta t)(1-\lambda_C\Delta t)(1-\lambda_h\Delta t)P_{n-1}(t) + (n-3)\lambda\Delta t P_3(t)$$
$$\frac{P_{n-1}(t+\Delta t) - P_{n-1}(t)}{\Delta t} + (\phi(x) + \lambda + \lambda_C + \lambda_h)P_{n-1}(t) = (n-3)\lambda P_3(t)$$

Now, taking $\lim\limits_{\Delta t \to 0}$, we get:

$$\lim_{\Delta t \to 0} \frac{P_{n-1}(t+\Delta t) - P_{n-1}(t)}{\Delta t} + (\phi(x) + \lambda + \lambda_C + \lambda_h)P_{n-1}(t) = (n-3)\lambda P_3(t)$$
$$\left[\frac{\partial}{\partial t} + \phi(x) + \lambda + \lambda_C + \lambda_h\right]P_{n-1}(t) = (n-3)\lambda P_3(t) \tag{5}$$

For state S_5,

$$P_S(t+\Delta t) = (1-\alpha\Delta t)(1-\delta(x)\Delta t)(1-\lambda_C\Delta t)(1-\lambda_h\Delta t)P_S(t) + \alpha\Delta t P_0(t) + \lambda\Delta t P_{n-1}(t)$$
$$\frac{P_S(t+\Delta t) - P_S(t)}{\Delta t} + (\alpha + \delta(x) + \lambda_C + \lambda_h)P_S(t) = \alpha P_0(t) + \lambda P_{n-1}(t)$$

Now, taking $\lim\limits_{\Delta t \to 0}$, we get:

$$\lim_{\Delta t \to 0} \frac{P_S(t+\Delta t) - P_S(t)}{\Delta t} + (\alpha + \delta(x) + \lambda_C + \lambda_h)P_S(t) = \alpha P_0(t) + \lambda P_{n-1}(t)$$
$$\left[\frac{\partial}{\partial t} + \alpha + \delta(x) + \lambda_C + \lambda_h\right]P_s(t) = \alpha P_0(t) + \lambda P_{n-1}(t) \tag{6}$$

For state S_6,

$$P(x+\Delta x, t+\Delta t) = \{1 - \delta(x)\Delta t\}P(x,t) \tag{7}$$

For state S_7/S_8

$$P_i(x+\Delta x, t+\Delta t) = \{1 - \phi_j(x)\Delta t\}P_i(x,t) \tag{8}$$

Boundary conditions of the system are obtained corresponding to transitions between the states where transition from a state with and without elapsed repair time exists when elapsed repair times x and 0. Hence, the following boundary conditions are obtained:

$$P(0,t) = \alpha P_s(t) \tag{9}$$

$$\begin{aligned} P_i(0,t) &= \lambda_i \sum_j P_j(t); \\ &\text{where } i = h, j = S,0,1,2,3,n-1 \\ &\qquad i = C, j = S,0,1,2,3,n-1 \text{ respectively} \end{aligned} \tag{10}$$

When system is perfectly good, i.e., in initial state S_0, then $P_0(0) = 1$ and all other state probabilities are zero at $t = 0$.

Taking the Laplace transformation from Eq. (1) to (10), one can get:

$$\begin{aligned} (s + n\lambda + \lambda_h + \lambda_C + \alpha)\overline{P}_0(s) &= 1 + \delta(x)\overline{P}_s(s) + \phi(x)\overline{P}_1(s) \\ &+ \int_0^\infty \phi_h(x)\overline{P}_h(x,s)\mathrm{d}x + \int_0^\infty \delta(x)\overline{P}(x,s)\mathrm{d}x + \int_0^\infty \phi_C(x)\overline{P}_C(x,s)\mathrm{d}x \end{aligned} \tag{11}$$

$$(s + \phi(x) + (n-1)\lambda + \lambda_C + \lambda_h)\overline{P}_1(s) = n\lambda\overline{P}_0(s) + \phi(x)\overline{P}_2(s) \tag{12}$$

$$(s + \phi(x) + (n-2)\lambda + \lambda_C + \lambda_h)\overline{P}_2(s) = (n-1)\lambda\overline{P}_1(s) + \phi(x)\overline{P}_3(s) \tag{13}$$

$$(s + \phi(x) + (n-3)\lambda + \lambda_C + \lambda_h)\overline{P}_3(s) = (n-2)\lambda\overline{P}_2(s) + \phi(x)\overline{P}_{n-1}(s) \tag{14}$$

$$(s + \phi(x) + \lambda + \lambda_C + \lambda_h)\overline{P}_{n-1}(s) = (n-3)\lambda\overline{P}_3(s) \tag{15}$$

$$(s + \alpha + \delta(x) + \lambda_C + \lambda_h)\overline{P}_s(s) = \alpha\overline{P}_0(s) + \lambda\overline{P}_{n-1}(s) \tag{16}$$

$$\left[\frac{\partial}{\partial x} + s + \delta(x)\right]\overline{P}(x,s) = 0 \tag{17}$$

$$\left[\frac{\partial}{\partial x} + s + \phi_j(x)\right]\overline{P}_i(x,s) = 0; \quad \text{where } i = h, C;\ j = h, C \text{ respectively} \tag{18}$$

$$\overline{P}(0,s) = \alpha\overline{P}_s(s) \tag{19}$$

$$\begin{aligned} \overline{P}_i(0,s) &= \lambda_i \sum_j \overline{P}_j(s); \\ &\text{where } i = h,\ j = S,0,1,2,3,n-1 \\ &\qquad i = C,\ j = S,0,1,2,3,n-1 \text{ respectively} \end{aligned} \tag{20}$$

Solving equations from (11) to (20) with the help of initial and boundary conditions, one can obtain

$$\overline{P}_0(s) = \frac{1}{(H_8 - H_5 - H_6 - H_7)} \tag{21}$$

$$\overline{P}_1(s) = \frac{\overline{P}_0(s) n\lambda}{\left[s + (n-1)\lambda + \lambda_C + \lambda_h + \phi(x) - \frac{\phi(x)(n-1)\lambda}{H_1}\right]} \tag{22}$$

$$\overline{P}_2(s) = \frac{\overline{P}_0(s) n(n-1)\lambda^2}{H_2} \tag{23}$$

$$\overline{P}_3(s) = \overline{P}_0(s) H_3 \tag{24}$$

$$\overline{P}_{n-1}(s) = P_0(s) H_4 \tag{25}$$

$$\overline{P}_s(s) = \frac{(\alpha + \lambda H_4) P_0(s)}{[s + \alpha + \lambda_C + \lambda_h + \delta(x)} \tag{26}$$

where

$$H_1 = s + (n-2)\lambda + \lambda_h + \lambda_C + \phi(x) - \frac{\phi(x)(n-2)\lambda}{\left[s + (n-3)\lambda + \lambda_C + \lambda_h + \phi(x) - \frac{(n-3)\phi(x)\lambda}{[s + \lambda + \lambda_C + \lambda_h + \phi(x)]}\right]}$$

$$H_2 = H_1[s + (n-1)\lambda + \lambda_C + \lambda_h + \phi(x)] - \phi(x)(n-1)\lambda$$

$$H_3 = \frac{n(n-1)(n-2)\lambda^3}{H_2\left[s + (n-3)\lambda + \lambda_C + \lambda_h + \phi(x) - \frac{(n-3)\phi(x)\lambda}{(s + \lambda + \lambda_C + \lambda_h + \phi(x))}\right]}$$

$$H_4 = \frac{H_3 \lambda (n-3)}{(s + \lambda + \lambda_C + \lambda_h + \phi(x))}$$

$$H_5 = \frac{n\lambda\left[\phi(x) + \frac{\lambda_h \phi_h(x)}{(s + \phi_h(x))} + \frac{\lambda_C \phi_C(x)}{(s + \phi_C(x))}\right]}{\left[s + (n-1)\lambda + \lambda_C + \lambda_h + \phi(x) - \frac{\phi(x)(n-1)\lambda}{H_1}\right]}$$

$$H_6 = (\alpha + \lambda H_4)\left[\delta(x) + \frac{\lambda_h \phi_h(x)}{(s + \phi_h(x))} + \frac{\lambda_C \phi_C(x)}{(s + \phi_C(x))} + \frac{\alpha \delta(x)}{(s + \delta(x))}\right]$$

$$H_7 = \left[\frac{\lambda_h \phi_h(x)}{(s + \phi_h(x))} + \frac{\lambda_C \phi_C(x)}{(s + \phi_C(x))}\right]\left[1 + \frac{n(n-1)\lambda^2}{H_2} + H_3 + H_4\right]$$

$$H_8 = (s + n\lambda + \lambda_C + \lambda_h + \alpha)$$

The Laplace transformation of the probabilities that the system is in the up (i.e., in good or degraded state) and the down (failed state) state at any time t is as follows:

$$\overline{P}_{\text{up}}(s) = \overline{P}_0(s) + \overline{P}_1(s) + \overline{P}_2(s) + \overline{P}_3(s) + \overline{P}_{n-1}(s) + \overline{P}_{\text{s}}(s) \quad (27)$$

$$\overline{P}_{\text{down}}(s) = \overline{P}_{\text{h}}(x,s) + \overline{P}(x,s) + \overline{P}_{\text{c}}(x,s) \quad (28)$$

4 Numerical Computations

4.1 Availability Analysis

Availability is the probability that a particular system is operating under a specified condition up to a specific time. The availability of a complex industrial system much more depends upon both the maintainability and reliability. To calculate the availability of the considered system, setting the different failure and repair rate as $\lambda = 0.025, \lambda_{\text{C}} = 0.001, \lambda_{\text{h}} = 0.03, \alpha = 0.015, \quad \phi(x) = \phi_{\text{C}}(x) = \phi_{\text{h}}(x) = \delta(x) = 1$ [28–30] with 500 parallel units in the system in Eq. (27) then taking the inverse Laplace transform, we get the availability of the system. Varying time unit t from 0 to 10 units of time in availability expression, one can obtain the following Fig. 2 which represents the behavior of availability for the system.

Figure 2 reflects that the availability of the considered system initially decrease rapidly and then smoothly as time passes as at t = 1 unit of time availability of the system is 0.98 then at t = 2 unit of time it becomes 0.972 and after this it slowly-slowly becomes approximately constant. This slowdown shows the importance of right maintenance policy.

Fig. 2 Availability as function of time

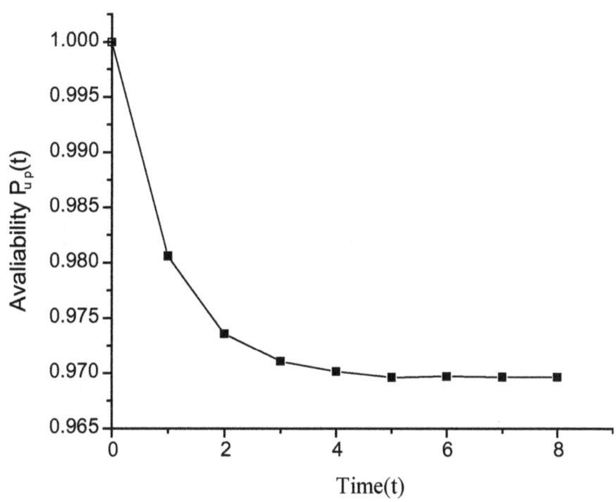

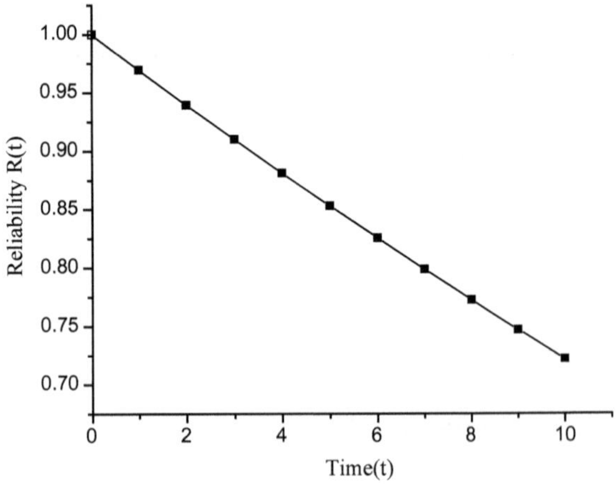

Fig. 3 Reliability as function of time

4.2 *Reliability Analysis*

The reliability of a system can be calculated by putting all repairs as zero [31, 32] because reliability did not consider maintenance in account, with same set of failure rates as discussed in Sect. 4.1 in Eq. (27). Now with the help of Laplace transform, one can obtain reliability of the considered system as:

$$R(t) = \begin{pmatrix} 292.7164134\,\mathrm{e}^{(-12.546t)} + 625\,\mathrm{e}^{(-12.526t)}\sin\mathrm{h}(0.02t) \\ -1.510280805\,\mathrm{e}^{(-0.056t)} - 115.7495976\,\mathrm{e}^{(-12.481t)} - 219.018\,\mathrm{e}^{(-12.506t)} \\ +42.05155976\,\mathrm{e}^{(-12.456t)} + 2.510257987\,\mathrm{e}^{(-0.046t)} \end{pmatrix} \tag{29}$$

Varying time unit t from 0 to 10 in the Eq. (29), the reliability of the system graph is obtained as shown in Fig. 3.

The above graph of reliability shows that it decreases in a straight line with respect to time unit t. The difference between the graph of reliability and availability shows the importance of maintenance in the system.

4.3 *Mean Time to Failure (MTTF) Analysis*

MTTF of a system plays a very important role to find about the behavior of failure rates of the systems. It is defined as the average failure time of the system during its operation. For MTTF evaluation of the system, consider all repairs as zero with

Table 2 MTTF as function of failure rates

Variations in failure rates λ, λ_C, λ_h, α	MTTF with respect to failure rates			
	λ	λ_C	λ_h	α
0.01	29.731480	22.405968	54.596143	28.776875
0.02	28.174278	18.482279	36.897200	26.655239
0.03	27.122119	15.702100	27.601529	25.224535
0.04	26.365150	13.634757	21.942996	24.192828
0.05	25.794668	12.040145	18.162099	23.412351
0.06	25.349379	10.774346	15.468309	22.800288
0.07	24.992170	9.746087	13.456958	22.306621
0.08	24.699268	8.894802	11.900592	21.899359
0.09	24.454746	8.178777	10.662020	21.557083
0.10	24.247534	7.568380	9.653803	21.264916

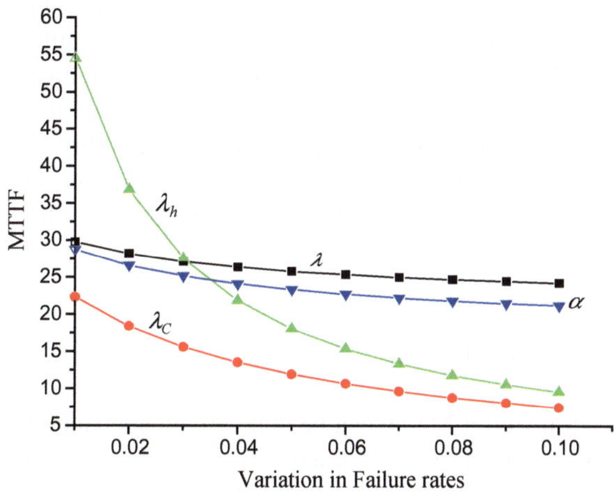

Fig. 4 MTTF as function of failure rates

$\lambda = 0.025, \lambda_C = 0.001, \lambda_h = 0.03, \alpha = 0.015$ in reliability expression and take the limit as s tends to zero. Now varying λ, λ_C, λ_h and α one by one from 0.01 to 0.09, the MTTF of the system is obtained as tabulated in Table 2 and its variation as shown in Fig. 4.

From Fig. 4, it is concluded that the MTTF of the system decreases with respect to all failures. It is observed that MTTF decaying rapidly with respect to human failure in comparison with other failures which reflects that the system fails more quickly by human failure as compared to other failures. MTTF have least variation with respect to unit failure which shows that the system is not often failed due to failure of any unit of the system.

4.4 Sensitivity Analysis

Sensitivity analysis is a process which is used to find out the compassion in various failure rates with respect to system reliability/MTTF. It is defined as the partial derivatives of a factor with respect to the input parameters. With the aid of this analysis, one can find out that which failure affects the system reliability/MTTF. Here in this chapter, this analysis is done by the authors for system reliability and MTTF.

4.4.1 Sensitivity of Reliability

The sensitivity analysis of system's reliability is carried out by differentiating the reliability expression with respect to input parameter [22, 33] that is to say failure rates of the system and then setting the value of various failure rates as $\lambda = 0.025, \lambda_C = 0.001, \lambda_h = 0.03, \alpha = 0.015$. Taking t = 0–10 units of time in the partial derivatives of reliability with respect to different failure rates, one can obtain the Table 3 and Fig. 5 for sensitivity of reliability respectively as:

From Fig. 5, it is observed that the sensitivity of reliability decreases as time passes which reflects that the system reliability is more sensitive with respect to human error and catastrophic failure. So, it is observed that the system can be made less sensitive with respect to reliability by controlling human errors and catastrophic failure rates.

4.4.2 MTTF Sensitivity

Sensitivity of MTTF can be done by differentiating MTTF expression with respect to different failure rates [20], and then putting the values of different failure rates as discussed

Table 3 Sensitivity of reliability as function of time

Time (t)	$\frac{\partial R(t)}{\partial \lambda}$	$\frac{\partial R(t)}{\partial \lambda_C}$	$\frac{\partial R(t)}{\partial \lambda_h}$	$\frac{\partial R(t)}{\partial \alpha}$
0	0	0	0	0
1	−0.005659	−0.969372	−0.969372	−0.007923
2	−0.025095	−1.878722	−1.878722	−0.036553
3	−0.055417	−2.729882	−2.729882	−0.084075
4	−0.094797	−3.524732	−3.524732	−0.147563
5	−0.141597	−4.265193	−4.265193	−0.224378
6	−0.194359	−4.953207	−4.953207	−0.312149
7	−0.251785	−5.590735	−5.590735	−0.408745
8	−0.312724	−6.179737	−6.179737	−0.512260
9	−0.376157	−6.722173	−6.722173	−0.620994
10	−0.441189	−7.219987	−7.219987	−0.733431

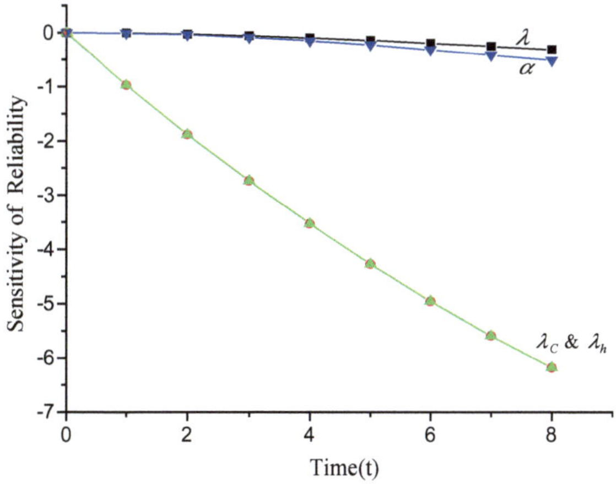

Fig. 5 Sensitivity of reliability as function of time

Table 4 Sensitivity of MTTF as function of failure rates

Variations in λ, λ_C, λ_h, α	$\frac{\partial(\text{MTTF})}{\partial\lambda}$	$\frac{\partial(\text{MTTF})}{\partial\lambda_C}$	$\frac{\partial(\text{MTTF})}{\partial\lambda_h}$	$\frac{\partial(\text{MTTF})}{\partial\alpha}$
0.01	−192.477648	−472.374439	−2548.059262	−263.561602
0.02	−125.680758	−325.649537	−1223.184168	−170.843426
0.03	−88.059865	−237.232681	−704.727656	−119.850275
0.04	−65.065249	−180.096418	−453.752154	−88.840124
0.05	−50.017850	−141.158008	−314.800730	−68.587561
0.06	−39.641820	−113.488873	−230.399072	−54.636069
0.07	−32.187623	−93.153596	−175.531560	−44.618827
0.08	−26.653490	−77.787880	−137.965970	−37.184298
0.09	−22.432654	−65.904445	−111.173601	−31.515144

in Sect. 4.4.1, the values of partial derivatives $\frac{\partial(\text{MTTF})}{\partial\lambda}, \frac{\partial(\text{MTTF})}{\partial\lambda_C}, \frac{\partial(\text{MTTF})}{\partial\lambda_h}, \frac{\partial(\text{MTTF})}{\partial\alpha}$ is obtained. Varying the failure rates one by one as beginning from 0.01 to 0.09, respectively, with interval 0.01, in these partial derivatives, one can obtain the Table 4 and Fig. 6 for sensitivity of MTTF of the system as:

Figure 6 shows that the sensitivity of MTTF increases with the increment in failure rates. Critical observation of the graph points out that MTTF of the system is more sensitive again with respect to human error.

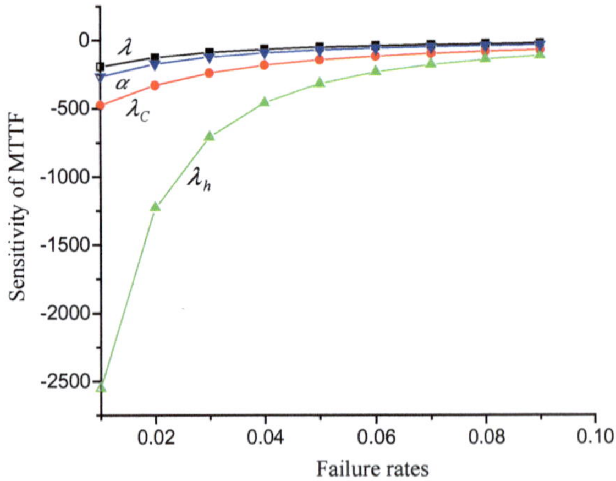

Fig. 6 Sensitivity of MTTF as function of failure rates

5 Conclusion

In this chapter, the authors investigated a complex industrial system to evaluate its various performance measures in order to enhance the system performance. It is concluded that a right maintenance policy is the key to improve a system performance (which is observed from the difference in graphs of system reliability and availability). Also it is observed that the system fails more quickly by human failure in comparison with other failures. Finally, it is observed that the considered system required more focus on human failure and catastrophic failure to reduce its sensitivity with respect to system reliability/MTTF for enhance its performance. It asserts that the result of this research will be useful for the management of such complex systems.

References

1. Gupta S, Tewari PC, Sharma AK (2009) A Markov model for performance evaluation of a coal handling unit of a thermal power plant. J Ind Syst Eng 3(2):85–96
2. Khanduja R, Tewari PC, Kumar D (2010) Mathematical modelling and performance optimization for the digesting system of a paper plant. Int J Eng 23(3&4):215–225
3. Burgio AD, Menniti AP, Sorrentino N (2007) The reliability evaluation of a power system in presence of photo-voltaic and wind power generation plants and UPS. In: Proceeding of 9th international conference on electrical power quality and utilisation IEEE, EPQU, University of Calabria, Rende, pp 1–6
4. Beamon BM (1998) Performance, reliability, and probability of material handling unit. Int J Prod Res 36(2):377–393
5. Castro HF, Cavalca K (2003) Availability optimization with genetic algorithm. Int J Qual Reliab Manage 20(7):847–863
6. Dhillon BS, Liu Y (2006) Human error in maintenance: a review. J Qual Maint Eng 12(1): 21–36

7. Kumar A, Ram M (2013) Reliability measures improvement and sensitivity analysis of a coal handling unit for thermal power plant. Int J Eng-Trans C 26(9):1059–1066
8. Distefano S (2011) The standby engineering: classification and quantification of standby in reliability. Int J Syst Assur Eng Manage 2(4):333–341
9. Zhang X, Pham H, Johnson CR (2010) Reliability models for systems with internal and external redundancy. Int J Syst Assur Eng Manage 1(4):362–369
10. Misra KB, Verma AK (2011) Special section on reliability and risk assessment of complex systems. IEEE Trans Reliab 60(1):59–60
11. Dhillon BS (1993) Reliability and availability analysis of a system with warm standby and common cause failure. Microelectron Reliab 33(9):1343–1349
12. Pan JN (1997) Reliability prediction of imperfect switching systems subject to multiple stresses. Microelectron Reliab 37(3):439–445
13. Azaron A, Katagiri H, Kato K, Masatoshi S (2006) Reliability evaluation of multi-component cold-standby redundant system. Appl Math Comput 173:137–149
14. Filieri A, Ghezzi C, Grassi V, Mirandola R (2010) Reliability analysis of component-based systems with multiple failure modes. Component-based software engineering. Springer, Berlin, pp 1–20
15. El-Said KM, El-Sherbeny MS (2010) Stochastic analysis of a two-unit cold standby system with two stage repair and waiting time. Indian J Stat 72-B(1):1–10
16. Ram M (2010) Reliability measures of a three-state complex system: a copula approach. Appl Appl Math Int J 5(10):1483–1492
17. Ram M, Singh SB (2010) Analysis of a complex system with common cause failure and two types of repair facilities with different distributions in failure. Int J Reliab Saf 4(4):381–392
18. Ram M, Singh SB, Singh VV (2013) Stochastic analysis of a standby system with waiting repair strategy. EEE Trans Syst Man Cybern Syst 43(3):698–707
19. Gupta S, Tewari PC (2011) Performance modelling of power generation system of a thermal plant. Int J Eng 24(3):239–248
20. Majeed AR, Sadiq NM (2006) Availability and reliability evaluation of Dokan hydro power station. In: IEEE PES transmission and distribution conference and exposition Latin America, Venezuela, pp 1–6
21. Ram M, Kumar A (2015) Paper mill plant performance evaluation with power supply in standby mode. Int J Qual Reliab Manage 32(4):400–414
22. Ram M, Kumar A (2014) Performance of a structure consisting a 2-out-of-3: F substructure under human failure. Arab J Sci Eng 39(11):8383–8394
23. Sawaragi T (1999) Modeling and analysis of human interaction with and within complex systems. In: IEEE international conference on system, men and cybernetics, Tokyo, vol 1, pp 701–707
24. Sutcliffe AG, Gregoriades A (2007) Automating scenario analysis of human and system reliability. IEEE Trans Syst Men Cybern-Part A Syst Hum 37(2):249–261
25. Weber P, Jouffe L (2006) Complex system reliability modelling with dynamic object oriented Bayesian networks (DOOBN). Reliab Eng Syst Saf 91(2):149–162
26. Ram M, Kumar A (2015) Performability analysis of a system under 1-out-of-2: G scheme with perfect reworking. J Braz Soc Mech Sci Eng 37(3):1029–1038
27. Ram M (2013) On system reliability approaches: a brief survey. Int J Syst Assur Eng Manage 4(2):101–117
28. Kumar A, Ram M (2015) Performance of marine power plant under generators, main switch board and distributed board failures. J Mar Sci Appl 14(4):450–458
29. Ram M, Singh SB (2009) Analysis of reliability characteristics of a complex engineering system under copula. J Reliab Stat Stud 2(1):91–102
30. Ram M, Singh SB (2012) Cost benefit analysis of a system under head-of-line repair approach using Gumbel-Hougaard family copula. J Reliab Stat Stud 5(2):105–118
31. Goyal N, Kaushik A, Ram M (2016) Automotive water cooling system analysis subject to time dependence and failure issues. Int J Manuf Mater Mech Eng 6(2):1–22

32. Manglik M, Ram M (2015) Behavioural analysis of a hydroelectric production power plant under reworking scheme. Int J Prod Res 53(2):648–664
33. Salehfar H, Trihadi S (1998) Application of perturbation analysis to sensitivity computations of generating units and system reliability. IEEE Trans Power Syst 13(1):152–158

An Insight into the Coanda Flow Through Mathematical Modeling

Maharshi Subhash, Michele Trancossi and José Páscoa

1 Introduction

The attachment of the jet over the adjacent curved surface was known around two centuries ago by Young [1] and patented around one century later by Henry Coanda, a Romanian engineer; therefore, this phenomena is known as 'Conada' effect [2]. The flow over curved surface has ample applications in engineering [3–12]. But, there is very few scientific understanding about it. In this research work, the Coanda flow is analyzed numerically. Its potential for future development of the cutting age technology can be seen in the project ACHEON (Aerial Coanda High Efficiency Orienting-jet Nozzle). This project describes about the thrust vectoring using the Coanda flow for the use in aviation for Short-Take-Off and Landing (STOL) which is supported in a patent developed at University of Modena and Reggio Emilia [13] called the HOMER (high-speed orienting momentum with enhanced reversibility) nozzle as shown in Fig. 1.

The detailed literature review in addition to its engineering applications specially for generating lift on the curved surface has been performed by Trancossi [14].

The mechanism of the flow over Coanda surface is described by Newman [15] found that flow adhesion is caused by the momentum balance between centrifugal force and the pressure force. Due to the interaction of the ambient fluid with the boundary layer, the static pressure increases gradually when the pressure gradient

M. Subhash (✉)
Department of Mechanical and Automobile Engineering, Graphic Era University, 566/6 Bell Road Clement town, Dehradun 248002, Uttarakhand, India
e-mail: maharshisubhash@gmail.com

M. Trancossi
Faculty of Arts, Engineering and Sciences, Sheffield Hallam University, City Campus Howard Street, Sheffield S1 1WB, UK

J. Páscoa
University of Beira Interior, Covilha, Portugal

M. Ram and J.P. Davim (eds.), *Modeling and Simulation in Industrial Engineering*, Management and Industrial Engineering, DOI 10.1007/978-3-319-60432-9_5

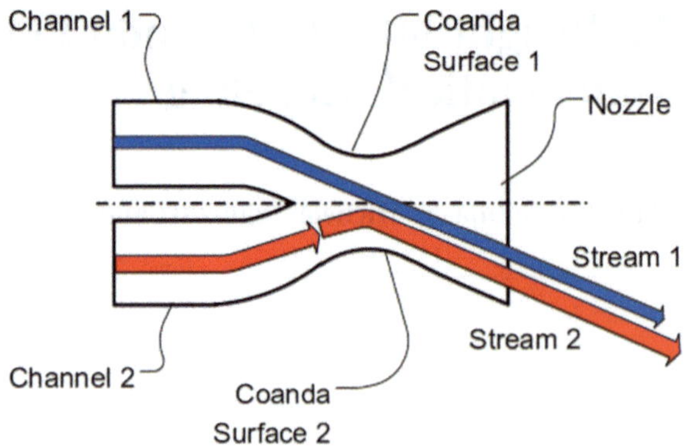

Fig. 1 Schematic of the Homer nozzle

becomes zero, that position on the curved surface is the verge of separation, beyond this position pressure gradient becomes positive causes the reverse flow. However, the above mechanism of flow adhesion is not complete in the sense that there exists two layer of velocity profile, first at the vicinity of the curved surface is influenced by viscous effect and another is by turbulent nature of the flow and also a abstruse phenomenon of the entrainment of the ambient fluid into the boundary layer causes the adhesion as well as separation of boundary layer [16]. Therefore, it is necessary to include the viscous term and advection term in the analysis. Hence, its complex flow behavior enhances the curiosity about the flow from mathematical point of view. Therefore, in this chapter, we attempted to clarify the mechanism of the flow over Coanda surface.

2 Problem Formulation

Let us consider a cylinder with radius R and the fluid jet with velocity U flows over the cylindrical surface from exit diameter d_j as shown in Fig. 2.

When the jet flows tangentially over the cylindrical surface, one cannot neglect viscous and inertia effect evolves more stiff equation to solve. Also, the previous research by Subhash and Dumas [17] shows that without resolving the viscous sublayer, one cannot get the numerically independent adhesion angle of the flow. When the flow propagates over the cylindrical surface tangentially, it incorporates two types of flow-free jet and wall shear flow. Hence, it introduces new type of flow, which needs careful study of the flow behavior at the boundary layer level, at which the wall shear stress affects the flow.

Here, an attempt has been made to understand the physics in more detail through numerical analysis.

Fig. 2 Flow over the curved surface (COANDA type flow)

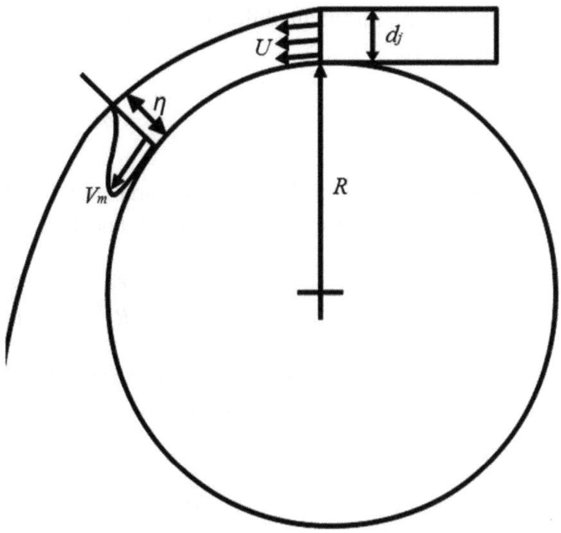

3 Mathematical Modeling

Consider the steady state two-dimensional and incompressible fluid flows over the cylindrical surface. The Navier–Stokes equations considering the boundary layer developed at the cylindrical surface are written in a curvilinear coordinate system. Assuming that the width of the jet slot is small compared to the radius of curvature of the cylinder, R.

Let us assume the tangential and radial velocity as follows.

$$V_\theta = u, V_r = v \tag{1}$$

Mass conservation and two-dimensional momentum conservation equations are written in Eqs. (2)–(4)

$$\frac{1}{r}\frac{\partial u}{\partial \theta} + \frac{\partial v}{\partial r} + \frac{v}{r} = 0 \tag{2}$$

$$v\frac{\partial u}{\partial r} + \frac{u}{r}\frac{\partial u}{\partial \theta} = v\frac{\partial^2 u}{\partial r^2} \tag{3}$$

$$\rho\frac{u^2}{r} = \frac{\partial p}{\partial r} \tag{4}$$

In the above equation, ν is defined as the laminar diffusivity and ν_t is the turbulent diffusivity and sum of these two $\nu + \nu_t = \nu_{\text{tot}}$ is defined as the total diffusivity.

The ratio of turbulent diffusivity to the laminar diffusivity is assumed in this form as shown in Eq. 5.

$$\frac{\nu_t}{\nu} = \sigma\theta^c \tag{5}$$

Introducing the non-dimensional parameter for independent parameter such as space and dependent parameter velocity and pressure.

$$u* = \frac{u}{U}, v* = \frac{v}{U}, r* = \frac{r}{R}, p* = \frac{p}{\rho U^2}, \quad \mathrm{Re} = \frac{UR}{\nu} \tag{6}$$

Substituting these non-dimensional parameters from Eq. (6) into Eqs. (2)–(4), we can get the following set of equations:

$$\frac{1}{r^*}\frac{\partial u^*}{\partial \theta} + \frac{\partial v^*}{\partial r^*} + \frac{v^*}{r^*} = 0 \tag{7}$$

$$v^*\frac{\partial u^*}{\partial r^*} + \frac{u^*}{r^*}\frac{\partial u^*}{\partial \theta} = \frac{1}{\mathrm{Re}}\frac{\partial^2 u^*}{\partial r^{*2}} \tag{8}$$

After non-dimensionalizing the governing equations, another step is to assume the stream function in such a way that the continuity equation should be satisfied and the assumed form is:

$$v* = -\frac{1}{r*}\frac{\partial \psi}{\partial \theta} \tag{10}$$

$$u^* = \frac{\partial \psi}{\partial \theta^*} \tag{11}$$

Furthermore, we can assume the linear equation of stream function as follows:

$$u^* = \theta^{a(c+1)} f(\eta) \tag{12}$$

$$\eta = \mathrm{Re}\, r * \frac{c+1}{\sigma}\theta^{(c+1)(a-1)} \tag{13}$$

Now, we can write the partial derivatives of stream function as ordinary differential equation as function of *f*.

$$\frac{\partial \psi}{\partial \theta} = a(c+1)\theta^{a(c+1)-1} f + \mathrm{Re} \cdot r * \cdot \frac{(c+1)^2}{\sigma}(a-1)\theta^{2(c+1)(a-1)} f' \tag{14}$$

$$\frac{\partial \psi}{\partial r*} = f' \cdot \theta^{(2a-1)(c+1)} \cdot \left(\frac{c+1}{\sigma}\right)\mathrm{Re} \tag{15}$$

$$\frac{\partial^2\psi}{\partial r*^2} = \mathrm{Re}^2 \cdot \left(\frac{c+1}{\sigma}\right)^2 \theta^{(3a-2)(c+1)} f'' \tag{16}$$

$$\frac{\partial^3\psi}{\partial r*^3} = \mathrm{Re}^3 \cdot \left(\frac{c+1}{\sigma}\right)^3 \theta^{(4a-3)(c+1)} f''' \tag{17}$$

$$\begin{aligned}\frac{\partial^2\psi}{\partial r\partial\theta} &= \mathrm{Re}\left(\frac{c+1}{\sigma}\right)(c+1)(2a-1)\theta^{(c+1)(2a-1)-1} f' \\ &\quad + \mathrm{Re}^2\left(\frac{c+1}{\sigma}\right)^2 r*(c+1)(a-1)\theta^{(c+1)(3a-2)-1} f''\end{aligned} \tag{18}$$

In terms of stream function, the momentum equation is written as:

$$-\frac{1}{r*}\frac{\partial\psi}{\partial\theta}\frac{\partial^2\psi}{\partial r*^2} + \frac{1}{r*}\frac{\partial\psi}{\partial r*}\frac{\partial^2\psi}{\partial r\partial\theta} = \frac{1}{\mathrm{Re}}\frac{\partial^3\psi}{\partial r*^3} \tag{19}$$

Elimination of the stream function with the variable f and η the equation takes the form:

$$\frac{\mathrm{d}^3 f}{\mathrm{d}\eta^3} + af\frac{\mathrm{d}^2 f}{\mathrm{d}\eta^2} + (1-2a)\left(\frac{\mathrm{d}f}{\mathrm{d}\eta}\right)^2 = 0 \tag{20}$$

This is the momentum equation as ordinary differential form and subsequent boundary conditions are as follows :

$$\eta = 0 : f = 0, \mathrm{d}f/\mathrm{d}\eta(0) = 0(\text{no} - \text{slip condition}) \tag{21}$$

$\eta \to \infty : \mathrm{d}f/\mathrm{d}\eta = [0 - D], \mathrm{d}^2 f/\mathrm{d}\eta^2 = 0$

(condition at the outer edge of boundary layer for all types of flow either laminar or turbulent.)

(22)

Such type of boundary condition is taken in order to incorporate the large range of flow, such as laminar to turbulent and/or dual jet system.

4 Solution Methodology

Now, rewriting Eq. (20) in the form of Cauchy problem:

$$\frac{\mathrm{d}^3 f}{\mathrm{d}\eta^3} = -af\frac{\mathrm{d}^2 f}{\mathrm{d}\eta^2} - (1-2a)\left(\frac{\mathrm{d}f}{\mathrm{d}\eta}\right)^2 \tag{23}$$

$$f(0) = 0, \mathrm{d}f/\mathrm{d}\eta(0) = 0, \mathrm{d}^2f/\mathrm{d}\eta^2(0) = b \tag{24}$$

where a and b are (positive) real numbers. The first equation is a scalar, autonomous, normal, third-order ordinary differential equation. It is equivalent to a system of first-order ordinary differential equations with given initial conditions.

$$\boldsymbol{f}' = \boldsymbol{H}(\boldsymbol{f}) \tag{25}$$

$$\boldsymbol{f}(0) = \boldsymbol{b} \tag{26}$$

where $\boldsymbol{b} \equiv (0, 0, b)$ while $\boldsymbol{f} \equiv (f_1, f_2, f_3)$ and $\boldsymbol{H} \equiv (H_1, H_2, H_3)$ are vector functions such that

$$f_1 = f, f_2 = \frac{\mathrm{d}f}{\mathrm{d}\eta}, f_3 = \frac{\mathrm{d}^2f}{\mathrm{d}\eta^2} \tag{27}$$

$$H_1(\boldsymbol{f}) = f_2, H_2(\boldsymbol{f}) = f_3, H_3(\boldsymbol{f}) = -af_1f_3 - (1-2a)f_1^2 \tag{28}$$

From (28), we see that $\boldsymbol{H}$ is a C^1 function, i.e., continuous with all its first derivatives. So it is *locally Lipschitz continuous*, and it is easy to verify that the Cauchy problem (23)–(24) admits a unique global solution in R [18]. Besides, since (23) is normal and the term on the right is given by a polynomial combination of lower-order derivatives, its solution is actually a C^∞ function.

In order to numerically solve Eq. (23), we applied the following version of the Taylor series method [19]. We consider the Taylor series truncated to the fifth order for f, to the fourth order for $\mathrm{d}f/\mathrm{d}\eta$ and third order for $\mathrm{d}^2f/\mathrm{d}\eta^2$. Then, if f and its first and second derivatives are known at a given point η, for the point $\eta + \delta\eta$ we have

$$\begin{aligned} f(\eta+\delta\eta) = {} & f(\eta) + f'(\eta)\delta\eta + \frac{1}{2}f''(\eta)(\delta\eta)^2 + \frac{1}{6}f'''(\eta)(\delta\eta)^3 + \frac{1}{24}f^{(iv)}(\eta)(\delta\eta)^4 \\ & + \frac{1}{120}f^{(v)}(\eta)(\delta\eta)^5 + o(\delta\eta)^6 \end{aligned} \tag{29}$$

$$\begin{aligned} f'(\eta+\delta\eta) = {} & f'(\eta) + f''(\eta)\delta\eta + \frac{1}{2}f'''(\eta)(\delta\eta)^2 + \frac{1}{6}f^{(iv)}(\eta)(\delta\eta)^3 \\ & + \frac{1}{24}f^{(v)}(\eta)(\delta\eta)^4 + o(\delta\eta)^5 \end{aligned} \tag{30}$$

$$f''(\eta+\delta\eta) = f''(\eta) + f'''(\eta)\delta\eta + \frac{1}{2}f^{(iv)}(\eta)(\delta\eta)^2 + \frac{1}{6}f^{(v)}(\eta)(\delta\eta)^3 + o(\delta\eta)^4 \tag{31}$$

The third derivative at point $\eta + \delta\eta$ is not given by a Taylor expansion, but it is directly calculated from the values of f, f' and f'', $\eta + \delta\eta$ at previously determined in

(29–31) and inserted in (23). In the same way, the higher-order derivatives are evaluated. Observe that from (23), we have

$$\frac{d^4f}{d\eta^4} = -af\frac{d^3f}{d\eta^3} - (2-3a)\frac{df}{d\eta}\frac{d^2f}{d\eta^2} \tag{32}$$

$$\frac{d^5f}{d\eta^5} = af\frac{d^4f}{d\eta^4} - 2(1-a)\frac{df}{d\eta}\frac{d^3f}{d\eta^3} - (2-3a)\left(\frac{d^2f}{d\eta^2}\right)^2 \tag{33}$$

Initial conditions at $\eta = 0$ are given by (21) for f, f' and f''. Initial conditions for f''', $f^{(iv)}$ and $f^{(v)}$ are deduced by inserting (24) in (23) and (28). We get

$$f'''(0) = 0, f^{(iv)}(0) = 0, f^{(v)}(0) = -(2-3a)b^2 \tag{34}$$

The maximum local error introduced in (27) is of order $O(\delta\eta)^4$. In our computations, we set $\delta\eta = 0.001$, so that for every iteration an error of $O(10^{-12})$ was introduced. Since we performed 10^6 iterations in order to investigate the asymptotic behavior of the functions up to $\eta_{max} = 10^6$. At $\delta\eta = 1000$, the global error was estimated to be $O(10^{-6})$.

In our computations, we considered different values for the parameter a in (23), while we kept fixed the value $b = 1$ in (24). There is actually no loss of generality in this approach, since Eq. (23) verifies *self-similarity* properties for which different solutions of (23)–(24) can be obtained from each other by means of simply scaling procedure.

In order to show this, let us assume $b \neq 0$ and let f a solution of the Cauchy problem (23)–(24). Let us now divide (23) for $b^{4/3}$, we get

$$\frac{d^3(f/b^{1/3})}{d(b^{1/3}\eta)^3} = -a\frac{f}{b^{1/3}}\frac{d^2(f/b^{1/3})}{d(b^{1/3}\eta)^2} - (1-2a)\left(\frac{d(f/b^{1/3})}{d(b^{1/3}\eta)}\right) \tag{35}$$

$$\frac{d^3g}{d\zeta^3} = -ag\frac{d^2g}{d\zeta^2} - (1-2a)\left(\frac{dg}{d\zeta}\right)^2 \tag{36}$$

where $g = f/b^{1/3}$ and $\zeta = b^{1/3}\eta$. So, g is also a solution of (23), but with initial conditions given by

$$g(0) = \frac{1}{b^{1/3}}f(0) = 0,\ g'(0) = \frac{1}{b^{2/3}}f'(0) = 0,\ g''(0) = \frac{1}{b}f''(0) = 1 \tag{37}$$

So the solution of the Cauchy problem (23)–(24) with $b = 0$ can be achieved by the solution f of (23)–(24) with a generic $b \neq 0$ simply by contracting f of a factor $1/b^{1/3}$ and the η variable of a factor $b^{1/3}$. This property can be easily seen in the following Fig. 3, in which the solutions of (23)–(24) for $b = 8$ and $b = 1$ for a fixed $a = 1/4$ are represented.

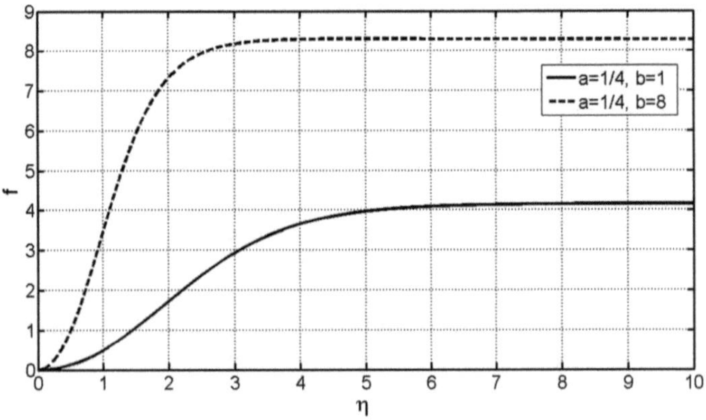

Fig. 3 Plot of streamline for the self-similar solution

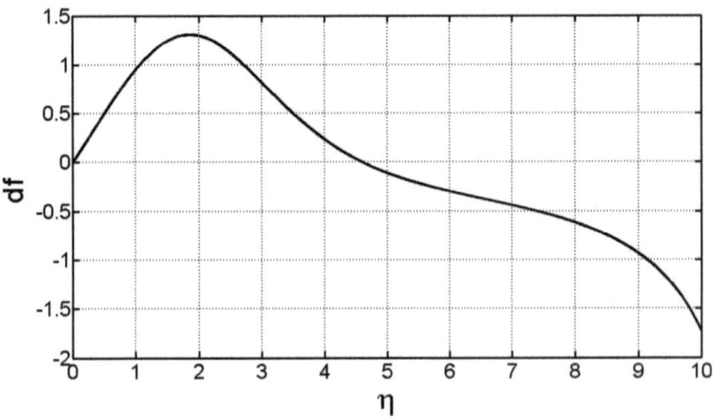

Fig. 4 Velocity profile for 'a = 0.20' and $d^2f/d\eta^2(0) = 1$

5 Results and Discussion

Feasible solution for certain range of parameter 'a' as shown in Eqs. (12), (13), and (20) is determined. Velocity profiles are plotted for various values of 'a' for the range [0.0–0.25], the plot is shown in Fig. 4, for 'a = 0.20' while keeping the parameter 'b = 1'. It can be said that for such range of 'a,' the velocity profile does not seem to be realistic.

Then, we checked for the range of value of 'a' [0.25–0.50]. All the velocity profile contained these range is a feasible solution of the equations as shown in Fig. 5.

The characteristics of the flow features will be different for the different values of 'a.' For the value of 'a = 0.25,' the value of velocity is zero for larger value of η.

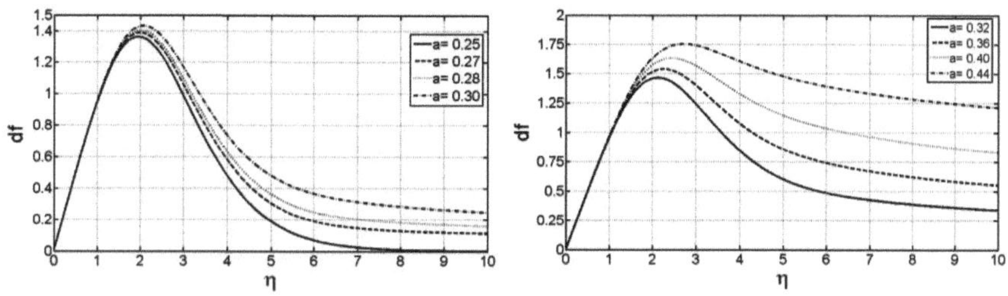

Fig. 5 Velocity profile for different values of 'a' keeping $d^2f/d\eta^2(0) = 1$

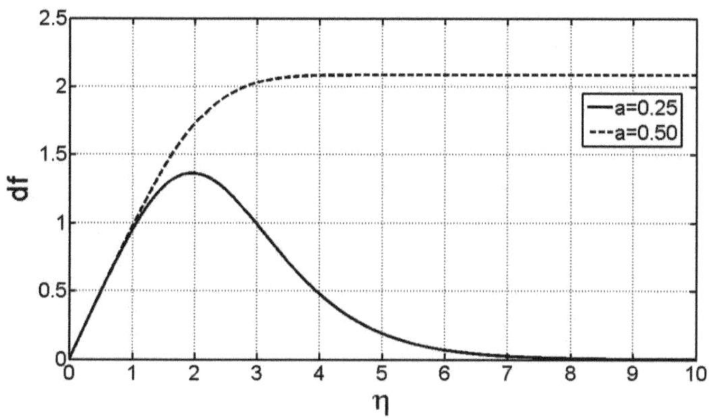

Fig. 6 Velocity profile keeping $d^2f/d\eta^2(0) = 1$, range of solution of the flow

This profile may be the case of single jet flowing over cylindrical surface with laminar flow characteristics.

At higher value of 'a,' the velocity does not fall to zero for any value of η, but the velocity gradient is zero for large value of η. Such profile may be the case of dual jet (velocity of the another jet is less than the jet adjacent to the cylindrical surface). However, qualitatively the dual jet exhibits the similar profile for '$a > 0.25$.'

Velocity profile is plotted in Fig. 6 for two extreme cases of '$a = 0.50$ and 0.25.' The outer bound of the solution is found at '$a = 0.50$.' Therefore, we can clearly define the solution range is [0.25–0.50] for different types of flow either single or dual jet depending upon the boundary condition of the equations. In case of dual jet flow, other parameters may also change, which will be discussed in the next section.

Now, in order to investigate the characteristic of velocity profile, the gradient of velocity (which is called as a double derivative of function f) against η is plotted as shown in Fig. 7. We see that two characteristics of velocity profile exist in the boundary layer of the flow over curved surface, which is different from the case of the flow over horizontal plane surface, which is also proved by some previous experiments [11] for such type of flow. The gradient of velocity profile is decreasing

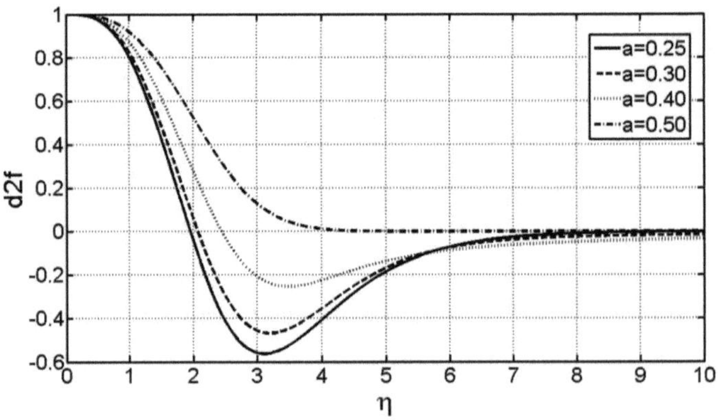

Fig. 7 Gradient of velocity profile on the cylindrical surface for the range of 'a = [0.25–0.5]'

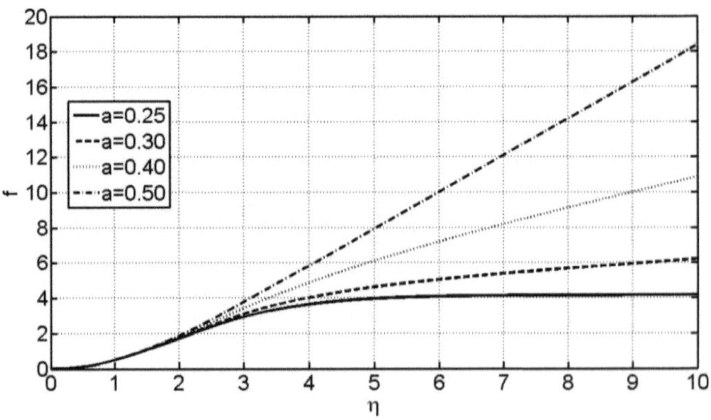

Fig. 8 Streamline curvature of the velocity profile at the cylindrical surface for $d^2f/d\eta^2(0) = 1$

until $\eta \approx 3$, then increasing until $\eta \approx 6$ and then remains invariant for the larger value of η. The decreasing gradient of the flow velocity is obvious due to the effect of wall shear stress at the vicinity of the cylindrical surface and its effect is until $\eta \approx 3$ (which is non-dimensional number of boundary-layer depth). It does not depend upon the flow characteristics either it is laminar or turbulent. It indicates that $\eta \approx 3$ may lie within the viscous sublayer. The rate of change of velocity gradient is dissipating in nature for '$a > 0.25$.' Here, one can see the effect of dual jet flow, on the gradient of velocity profile, but the extent of dissipating nature of velocity gradient is same until $\eta \approx 3$, which is described as the effect of the viscous sublayer.

Now, let us see the streamline within the boundary layer. The nature of streamline within the viscous sublayer signifies the invariant nature with respect to the types of flow and also invariant with initial value of $d^2f/d\eta^2(0) = b$ which can be seen in Figs. 8, 9 and 10. After the viscous sublayer, the direction of the flow is

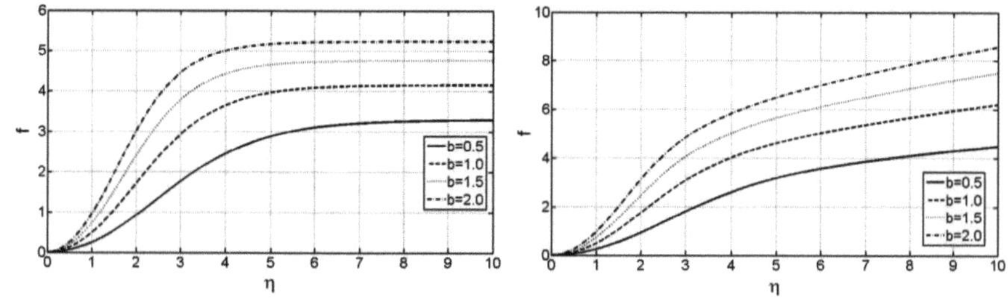

Fig. 9 Streamline curvature of the velocity profile at the cylindrical surface for 'a = 0.25' (*Left plot*) and 'a = 0.30' (*Right plot*)

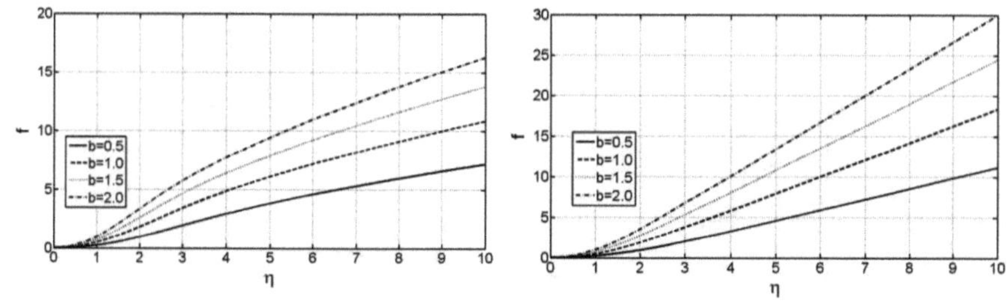

Fig. 10 Streamline curvature of the velocity profile at the cylindrical surface for 'a = 0.40' (*Left plot*), 'a = 0.50' (*Right plot*)

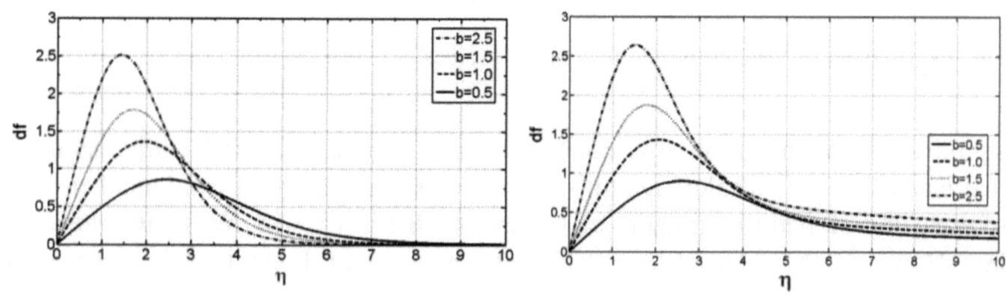

Fig. 11 Velocity profile for different values of $d^2f/d\eta^2(0) = b$, keeping 'a = 0.25' (*Left plot*), 'a = 0.30' (*Right plot*)

highly dependent upon the nature and intensity of the turbulent flow or type of the flow. At 'a = 0.25,' the flow direction is invariant with η at larger value. Therefore, number of turbulent length scale will be less. But for higher value of 'a,' the turbulent length scales will be enormous. Therefore, for values 'a > 0.25,' one needs to define all components of the Reynolds shear stress for the analysis. Which is not possible in the current analytical treatment of the flow.

Now, let us see the nature of velocity profile and flow behavior if we varied $d^2f/d\eta^2(0)$. Therefore, we plotted for $d^2f/d\eta^2(0) = b = 0.5, 1.00, 1.5, 2.5$ in Figs. 11 and 12. We see that the maximum velocity shifts to lower value of η. This

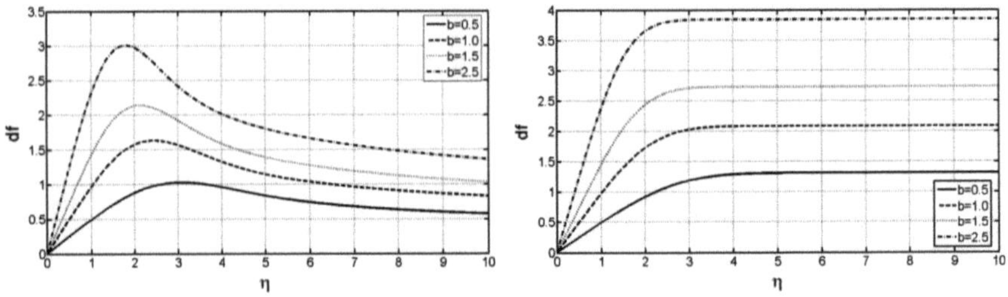

Fig. 12 Velocity profile for different values of $d^2f/d\eta^2(0) = b$, keeping '$a = 0.40$' (*Left plot*), '$a = 0.50$' (*Right plot*)

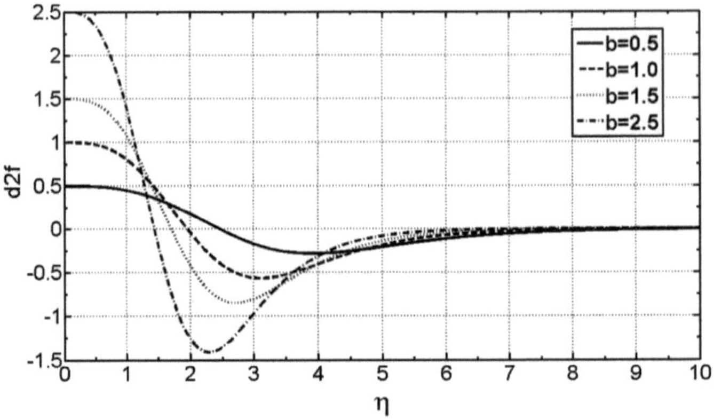

Fig. 13 Gradient of velocity profile for different values of $d^2f/d\eta^2(0) = b$, keeping '$a = 0.25$'

signifies that the thickness of viscous sublayer is decreasing, which further signifies that increasing the turbulent length scales. We examine this fact from another point of view and plotted the gradient of velocity profile in Fig. 13. We found the same fact. However, the value of '$a = 0.25$' is taken constant. It can be said that we can add the turbulent length scale by increasing the initial value of $d^2f/d\eta^2(0) = b$. In this way, flow can transit from laminar to turbulent. But, the determination of threshold value of transition can be the task of experimental investigation.

In this way, we have determined the two ways of increasing the turbulent length scale, first is altering the value of 'a,' and second way is varying the initial condition of $d^2f/d\eta^2(0)$, which further needs to be verified through experiment. One can employ both cases for the flow to be turbulent. As the length scales are not additive in nature, because of the nonlinear nature of the turbulent flow. One needs to evaluate different components of the Reynold's stress for the analysis.

The non-dimensional velocity at particular angle of the flow can be determined through Eqs. (38)–(39).

$$v* \approx -\theta^{a-1}[a \cdot f + (a-1)\eta f'] \tag{38}$$

$$u* \approx \mathrm{Re}\theta^{(2a-1)}f' \tag{39}$$

For the turbulent flow the value of $c = 1/3$, $\sigma = 1$, we want to calculate the turbulent diffusivity as shown in below:

$$v_t = 3.6439\ Kv\mathrm{Re}\theta^{1/2} \tag{40}$$

6 Conclusion

The mechanism of the Coanda flow is investigated. One should not neglect the viscous effect as it also affects the boundary layer. The effect of viscous is discussed in detail at the vicinity of the curved surface. Incorporation of set of boundary conditions gave the set of solutions which can be the case of dual jet and or turbulent jet flow. A nature of turbulent length scales was discussed and can be the source of the subgrid scale modeling, which will be performed in the next level of this research.

Acknowledgements The present work has been performed as part of the ACHEON Project | ACHEON Project—Aerial Coanda High Efficiency Orienting-jet Nozzle project, with ref. 309041 supported by the European Union through the Seventh Framework Program. The first author is highly indebted to Department of Science and methods in engineering. University of Modena and Reggio Emilia, Italy for the support of this work. The first author also would like to thanks to Dr. Andrea Andrisani for constructive advice on numerical computation.

References

1. Young T (1800) Outlines of experiments and inquires respecting sound and light. Phil Trans R Soc London 90:106–150
2. Coanda H (1936) Device for deflecting a stream of elastic fluid projected into an elastic fluid. US Patent # 2,052,869
3. Freund JB, Mungal MG (1994) Drag and wake modification of axisymmetric bluff bodies using Coanda blowing. J Aircr 31(3):572–578
4. Chang TL, Rachman A, Tsai HM, Zha Ge-C (2009) Flow control of an airfoil via injection and suction. J Aircr 46(1):291–300
5. Lee D-W, Hwang J-G, Kwon Y-D, Kwon S-B, Kim G-Y, Lee D-E (2007) A study on the air knife flow with Coanda effect. J Mech Sci Technol 21:2214–2220
6. Lalli F, Brusch A, Lama R, Liberti L, Mandrone S, Pesarino V (2010) Coanda effects in coastal flows. Coast Eng 57:278–289
7. Mabey K, Smith B, Whichard G, McKechnie T (2011) Coanda-assisted spray manipulation collar for a commercial plasma spray gun. J Therm Spray Technol 20(4):782–790
8. Kim H, Rajesh G, Setoguchi T, Matsuo S (2006) Optimization study of a Coanda ejector. J Therm Sci 15(4):331–336

9. Trancossi M, Madonia M, Dumas A et al (2016) A new aircraft architecture based on the ACHEON Coanda effect nozzle: flight model and energy evaluation. Eur Transp Res Rev 8:11. doi:10.1007/s12544-016-0198-4
10. Trancossi M, Subhash M, Angeli D (2013) Mathematical modelling of a two streams Coanda effect nozzle. ASME international mechanical engineering conference and exhibition, paper no. IMECE2013-63459
11. Dumas A, Subhash M, Trancossi M, Marques JP (2014) The influence of surface temperature on Coanda effect. Energy Proc 45:626–634
12. Trancossi M, Stewart J, Subhash M, Angeli D (2016) Mathematical model of a constructal Coanda effect nozzle. J Appl Fluid Mech 9(6):2813–2822
13. Trancossi M, Dumas A, Giuliani I, Baffigi I (2011) Ugello capace di deviare in modo dinamico e con-trollabile un getto sintetico senza parti meccaniche in movimento e suo sistema di controllo. Patent No. RE2011A000049, Italy
14. Trancossi M (2011) An overview of scientific and technical literature on Coanda effect applied to nozzles. SAE Technical Papers No. 2011-01-2591. ISSN: 0148-7191
15. Newman BG (1961) The deflexion of plane jets by adjacent boundaries, in Coanda effect. In: Lachmann GV (ed) Boundary layer and flow control. Pergamon Press, Oxford, vol 1, pp 232–264
16. Neuendorf R, Wygnansky I (1999) On a turbulent wall jet flowing over a circular cylinder. J Fluid Mech 381:1–25
17. Subhash M, Dumas A (2013) Computational study of Coanda adhesion over curved surface. SAE Int J Aerosp 6(1):260–272. doi:10.4271/2013-01-2302
18. Hartman P (1964) Ordinary differential equation, 1st edn. Wiley New York
19. Kincaid DR, Cheney EW (2002) Numerical analysis: mathematics of scientific computing, 3rd edn. American Mathematical Society

Preliminary CDF Assessment of an Innovative Propelled Wing with Enhanced Performances by Coanda Effect

Michele Trancossi and Maharshi Subhash

1 Introduction

One of the key issues of aeronautics has always been the solution to the problem of low-speed flight. Two solutions have reached an adequate success. They are helicopters and autogiros both are vertical axis propelled air vehicles. With the development of aviation technology, helicopter has coupled effective performances and flexibility [1–3], but presents lower energy efficiency with respect to any other air vehicle [4]. Helicopters are capable of vertical takeoff and landing (VTOL), vertical movements, forward and reverse flight, and hovering. This chapter presents a radically new solution that aims to realize radically new performances. Including VTOL operations, and hovering by means of a radically new and simple high-camber, blown wing by means of a novel ducted fan architecture with a dual use one based on Coanda effect and over-circulation and another specifically defined for VTOL operations. The novel wing is the definition of a new category of aircraft that can emulate operatively the tilt rotors such as Boeing Osprey [5, 6]. It has much lower complexity and higher robustness that allow mitigating the problem of instability during turning the motor propellers groups and lower efficiency against helicopters in vertical flight and against aircraft in horizontal flight [7]. In particular, the Phase II report on Osprey [7] demonstrates that twin-engine commercial tilt rotors derived from the military V-22 would be comparable to a medium-size commuter turboprop and could carry around 40 passengers up to

M. Trancossi (✉)
Faculty of Arts, Engineering and Sciences, Sheffield Hallam University,
City Campus Howard Street, Sheffield S1 1WB, UK
e-mail: mtrancossi@gmail.com

M. Subhash
Department of Mechanical and Automobile Engineering, Graphic Era University,
566/6 Bell Road Clement Town, Dehradun 248002, India
e-mail: maharshisubhash@gmail.com

M. Ram and J.P. Davim (eds.), *Modeling and Simulation in Industrial Engineering*,
Management and Industrial Engineering, DOI 10.1007/978-3-319-60432-9_6

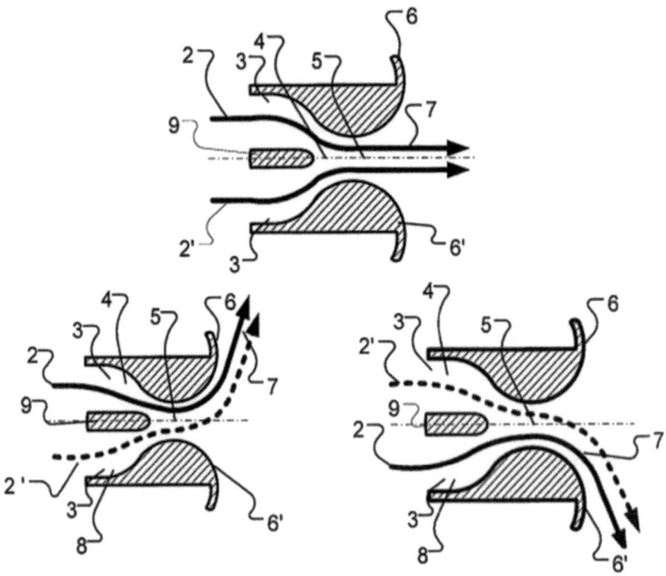

Fig. 1 ACHEON nozzle

600 miles. It would be capable of vertical takeoff and landing, but short takeoff and landing (STOL) rolls would improve payload or range. Tilt rotors could use existing airports, but 40-seat versions would be too large for many heliports. To compete with airline shuttle service, passenger cabin noise, vibration, and overall comfort levels will have to be at least equivalent to that of commuter aircraft, such as the DeHavilland Dash 8–300. Compared with turboprop aircraft, a tilt rotor would cost around 40–45% more at production level and about 14–18% more at operation level (over a 350-km trip). The proposed configuration has been possible starting by the results of the activity performed in the former EU FP7 ACHEON (Aerial Coanda High Efficiency Orienting Nozzle) Project [8, 9] which has demonstrated the possibility of deviating a propulsive synthetic jet, which is generated by two impinging streams, by means of Coanda effect (Fig. 1). The core of the ACHEON thrust and vector propulsion is a nozzle with a duct eventually bipartite into two internal channels, which converge in a single outlet with two facing Coanda surfaces (3) and (3').

Two impinging jets (2) and (2') generate a synthetic jet that proceeds straight if the streams have equal momentum, or adheres to the Coanda surface on the side of the stream with higher momentum. The direction of the synthetic jet depends on the momentums of the two jets. Control and stability are increased by Dielectric Barrier Discharge (DBD) installation [10, 11]. The architecture fits with electric propulsion and subsonic aircrafts (Mach 0–0.5). In particular, it has verified that the deflection angle of the jet (and of the thrust) is a function of both momentums (and speeds) of

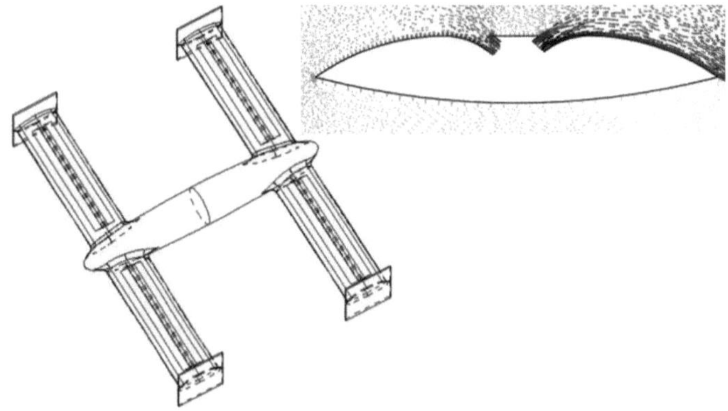

Fig. 2 Preliminary conceptual design of the aircraft [16]

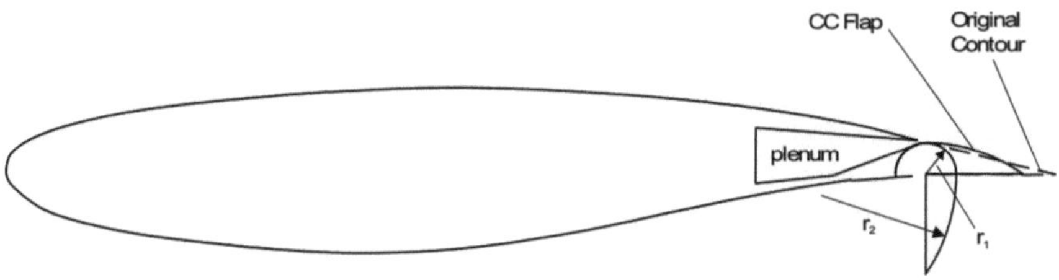

Fig. 3 Radelspiel dual curvature flapped wing

the two primitive streams and the geometric configuration of the nozzle [12, 13]. Trancossi et al. [14, 15] have clearly demonstrated the advantages of ACHEON in case of application to STOL aircraft both in terms of takeoff roll and in case of horizontal flight. ACHEON studies has allowed producing a new propelled wing configuration [16] and produced an effective energy model of a drone based on that wing (Fig. 2).

The second inspiration is the propelled wing with jet discharge studied by Radelspiel [17]. It shows that two strategies of blowing deserve particular attention (Fig. 3). The first strategy makes use of tangential blowing of thin high speed jets on the wall of a Coanda surface to overcome the adverse pressure gradients. This approach allows exploiting the high potentials of Coanda effect on fluid adhesion. The second strategy employs oblique blowing of air jets designed to generate longitudinal vortices in the boundary layer. The longitudinal vortices provide convective redistribution of momentum in the boundary layer, and they also enhance turbulent momentum transport. Landa et al. [18] have also studied the influence of stream-wise vortices on a generic high-lift configuration.

Some attempts of increasing the lift are connected to the increase of the lift by means of diffused propulsion embedded into the trailing of the wing. The most successful sample in this direction is constituted by the NASA LEAPtech project

Fig. 4 NASA LEAPtech wing during tests

[19, 20]. The project is indented to increase the lift coefficient of an aircraft wing by means of diffused electric propulsion. During preliminary testing on board of a truck, lift coefficients C_L in the range 5–7 have been achieved (Fig. 4).

Even if related to turbomachinery, another important direction of development that must be cited into the reference study on which the defined wing configuration is based relates to the huge study on Coanda effect by Drăgan. In particular, he has made a huge study on classical models of Coanda Effect [21] assessing the robustness of the Coanda effect model by Roderick [22] and Benner [23]. Drăgan [24] has also produced a fundamental reference regarding coupling super circulation with Coanda effect adhesion on rotary wing propulsion.

2 The New Propelled Wing Concept

The proposed propelled wing concept is mainly inspired by Benner's study [23] and Trancossi's blowing wing concept [16]. It is based on a dual system based on an unconventional high lift wing coupled with an electric ducted fan unit blowing on the top surface of the wing (Fig. 5).

Using Benner's model, it can be possible to describe Coanda adhesion in terms of equilibrium between centrifugal and pressure forces:

$$F_c = \frac{\rho \cdot R \cdot \mathrm{d}\theta \cdot \mathrm{d}Ru^2}{R} \tag{1}$$

$$F_p = R \cdot \mathrm{d}\theta \cdot \mathrm{d}p \tag{2}$$

where $d\theta$ is the infinitesimal angular element.

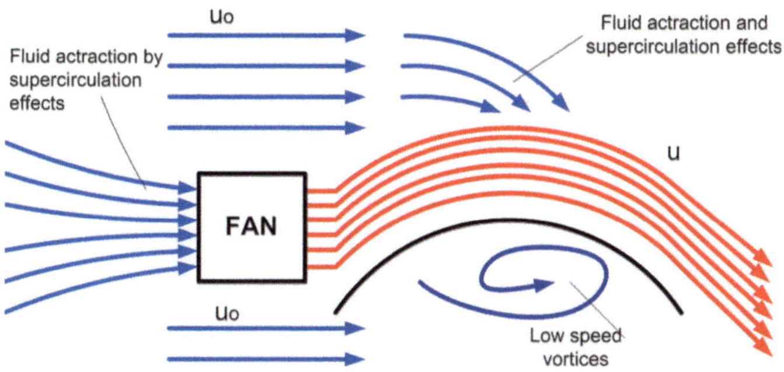

Fig. 5 Conceptual schematic of the proposed wing concept (*showing in red*) expected effect of Coanda adhesion

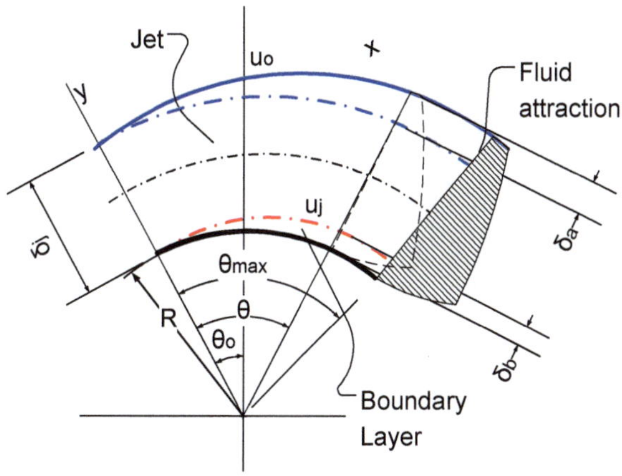

Fig. 6 Structure of the fluid dynamic jet

The model of the system can be subdivided into two control volumes according to Fig. 6. The first one is dominated by fluid deviation by means of the presence of the interaction of the jet through a surface. The second one is dominated by the Coanda adhesion caused the pressure phenomena.

Pressure drop along the jet can be expressed according to Benner [23] who present a crude but functional model for *h/R* ratios smaller than one. He produces a balance between the pressure forces and the centrifugal forces acting upon a volume of fluid.

$$y: p_{\text{staticjet}} - p_{\text{atm}} = -\frac{\rho \cdot u^2 \cdot h}{R} \tag{3}$$

Benner equation is based on difference in pressure only and works properly only for the condition $h/R < 1$ and does not consider the shear stress and the external fluid attraction. Dragan has verified the robustness of this formulation but has also evidenced the necessity of introducing a correction to allow Benner equation to fit better numerical results.

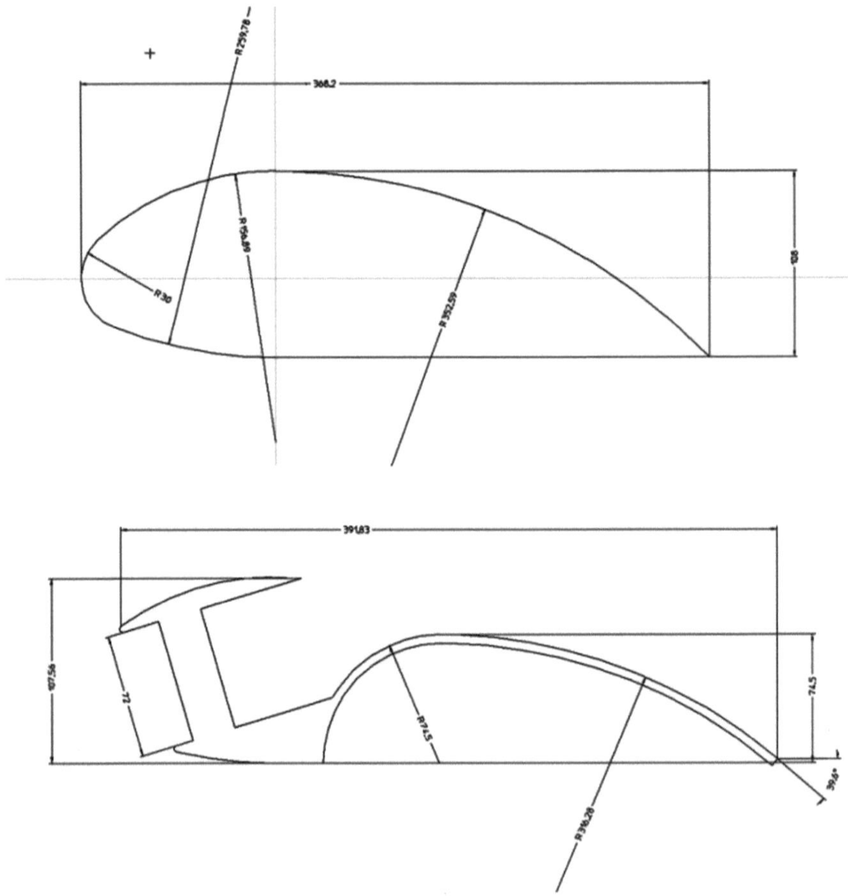

Fig. 7 Wing profiles with dimensions

The present research, on the other side, focuses on the definition of a more effective model, coupling it with the precedent models by Trancossi et al. [12]. The aim is not the production of any corrective coefficient that can fit the results. It is on the contrary the definition of a robust theoretical model that can fit a set of problems that have been marginally studied. In particular, the proposed area of interest focus in a large jet that blows almost tangentially on a convex surface.

This preliminary definition of a new wing concept assumes that R is large and some considerations can be done about the fluid jet development, and it can be considered that the development of the boundary layer is similar to the one over a flat plate. In particular, it can be assumed that the thickness δ of the jet is given to two different thicknesses: δ_b that refers to the boundary layer being caused by shear stress and to δ_a that refers to the jet and is caused by the phenomena of fluid attraction from the surrounding free stream.

3 Preliminary Model

It has been considered an initial 2D model. The following assumptions have been done:

1. The length over the Coanda surface has been indicated with x, $x = R\theta$ where θ is the generic angle of contact.
2. The Coanda jet considering the effects involved has been considered as the combination of three effects:
   - the boundary layer effect governed by shear stress;
   - the equilibrium of centrifugal and pressure forces (main Coanda effect);
   - the attraction of the surrounding slower fluid;

   The three effects can be considered separately and their effects can be added by superposition. In the core zone of the fluid, over the boundary layer boundary, the velocity is kept at least in one point equal to the initial velocity of the jet;
3. The jet has an initial constant profile of velocity.

The gauge pressure at the nozzle outlet is zero (i.e., $p_2 = 0$), since air is discharged to the atmosphere. According to the principle of conservation of mass, neglecting the fluid attraction phenomena, it results:

$$A_{\text{jet}} \cdot u_{\text{jet}} = A \cdot u \rightarrow u = \frac{A_{\text{jet}}}{A} \cdot u_{\text{jet}} \tag{4}$$

By assuming $x = \theta \cdot R$, and assuming, that R is sufficiently large, it has preliminary modeled the boundary layer according to the model which is used for the boundary layer on a flat plane. For the boundary layer term, it can be possible to consider the solution by Schlichting [25] in a turbulent regime:

$$Re_x = \frac{\rho \cdot u \cdot x}{\mu} \rightarrow \delta_b = 0.37 \frac{x}{\sqrt{Re_x}} \tag{5}$$

$$Re_x = \frac{\rho \cdot u \cdot x}{\mu} = \frac{\rho \cdot u \cdot \theta \cdot R}{\mu} = \theta \cdot Re_R \tag{6}$$

Skin friction for turbulent flows is given by

$$C_f = \frac{0.0594}{Re_x^{0.2}} = \frac{0.0594}{(\theta \cdot Re_R)^{0.2}} \tag{7}$$

Wall shear stress is another parameter of interest in boundary layers. It is usually expressed as a function of skin friction defined as:

$$C_f = \frac{2\tau}{\rho \cdot u^2} \rightarrow \tau = \frac{1}{2} \cdot C_f \cdot \rho \cdot u^2 = \frac{1}{2} \cdot \frac{0.0594}{(\theta \cdot Re_R)^{0.2}} \cdot Re_R^2 \cdot \frac{\mu^2}{\rho \cdot R^2} \tag{8}$$

and

$$\delta_b = 0.37 \frac{x}{\sqrt{Re_x}} \cong 0.37 \frac{\theta \cdot Re_R}{(\theta \cdot Re_R)^{0.5}} \cdot \frac{\mu}{\rho \cdot u} = 0.37 \cdot (\theta \cdot Re_R)^{0.5} \cdot \frac{\mu}{\rho \cdot u} \tag{9}$$

The fluid attraction phenomena can be modeled by the interaction of the external fluid and the jet. It is consequent of the difference between the static pressures of the surrounding fluid and the Coanda jet. It generates a reduction in the average velocity of the fluid, but generates an effective augmentation of the mass flow. This phenomenon presents more difficulties in terms of an effective modeling. It has been preliminarily neglected.

4 Thrust and Lift Behavior

The actual design problem relates in particular to an aircraft application. It is then necessary to produce a large CFD application that relates to different configurations. They are in particular a ultra high thickness wing, a configuration with ultra high lift and a high angle of deflection of the propulsive jet, a configuration with fluid deviation by means of the interior surface of the mobile wing, which will not considered into this specific chapter.

The objective is to verify if the above defined propelled wing could be the propelling element of a new aircraft system specialized for low-speed flight.

5 Numerical Activity

For each of the proposed wing models, it has been ensured the stability of the mesh by an effective CFD analysis. Two different calculations have been produced for each of section by using Ansys Fluent 17.1 [26] and EasyCFD [27]. For both of the model, an accurate grid independence analysis has been performed according to the ERCOFTAC [28] guidelines and as described in the following papers [29, 30]. The optimum number of grid (numerically stable grid) has been determined through the numerical computation of the grid at different refinement levels of the grid at the curved surface. It has been found that, when the grid resolved the viscous sub-layer until y + value less than two, then one can get the jet deflection angle independent of the grid. Preliminary simulations have been performed by using Spalart Allmaras [31] turbulence model and then SST K-ω [32, 33] model for more accurate calculations has been used on Ansys Fluent, and SST K-ω on Easy CFD been adopted.

Second-order upwind scheme has been used for discretizing the momentum equation and that of k and ω model.

With respect to the simulations through Ansys Fluent, the pressure and velocity have been coupled through the PISO (Pressure-Implicit with Splitting of Operators) method [34]. Pressure gradient term has been discretized using the PRESTO (PREssure STaggering Option) [35] method.

The PRESTO scheme provides improved pressure interpolation in situations where large body forces or strong pressure variation is present. Unsteady term has been discretized using first-order implicit method taking advantage of unconditionally stable with respect to time step size, which has been as taken $\Delta t = 1 \times 10^{-03}$ s. The three different systems have been computed with minor results between the two systems numerical codes with minor differences (less than 3%).

Final meshes have been generated by assuming the following parameters on the lines of the 2D profile:

1. lines discretization: 2.5×10^{-3} m;
2. inflation layers: height: 0.25×10^{-3}; expansion factor: 1.1; total levels: 10; max aspect ratio: 1/5.

A sample of the mesh of the wing is provided in Fig. 8. In particular, the following parameters have been reached:

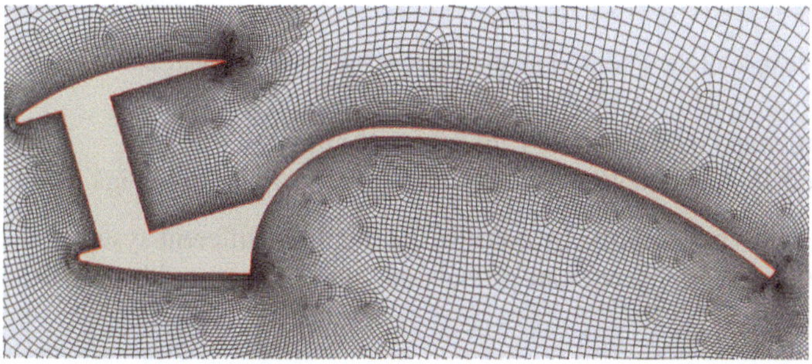

Fig. 8 Sample of the global mesh (EasyCFD)

Fig. 9 Skewness analysis of the mesh around a wing profile (max visible skewness 1.5 *green* cells in the visualized area)

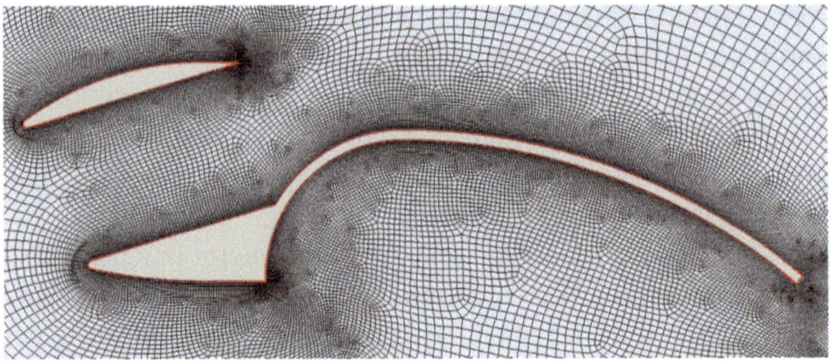

Fig. 10 Mesh of the wing without any source. (Easy CFD)

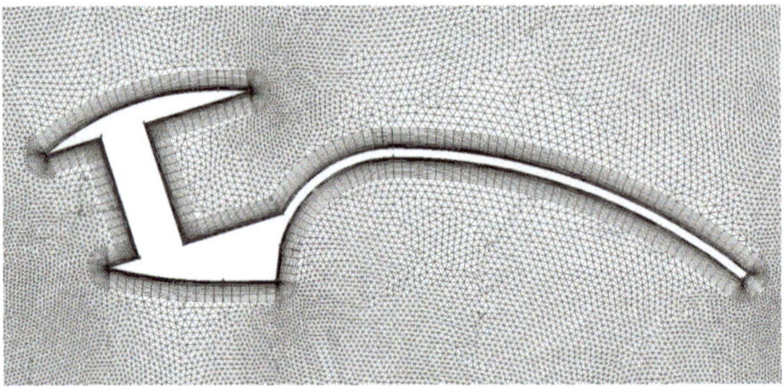

Fig. 11 Example of the Fluent tri-mesh with boundary layer refinement

1. **Fluent**: max skewness: 1.54, max aspect ratio: 9.11
2. **EasyCFD**: max skewness: 1.88; max aspect ratio 10.34 (Figs. 9–11)

Simulations have started by considering the three different systems: the full wing reference profile, wing without propulsion, and the wing with propulsion.

6 Results and Discussions

The simulations have been performed at different airspeeds in the range u_o = 0–20 m/s and different jet speeds $u_j = u_o + 5\text{–}u_o + 30$ m/s. In this case, different situation has been evaluated. The reference wing has been also evaluated in the same range. Results between Ansys Fluent and EasyCFD have pretty near (max error 5% in terms of generated lift) because of the accuracy of the mesh realized in both systems. Results for Ansys calculations have been reported in Figs. 12 (Lift force) and 13 (Drag coefficient).

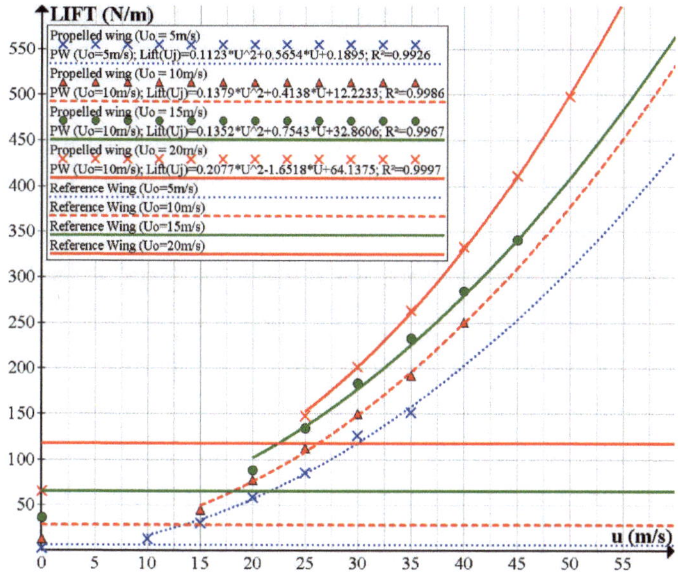

Fig. 12 Lift results (*lines* are CFD results and *symbols* are model results)

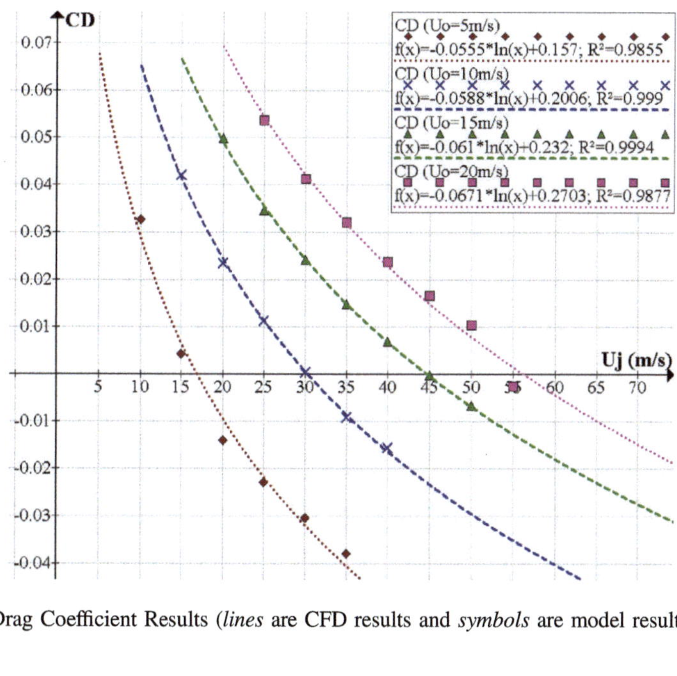

Fig. 13 Drag Coefficient Results (*lines* are CFD results and *symbols* are model results)

In Fig. 12, the horizontal axis is represented by jet velocity and vertical axis is represented by lift force for different air velocities. As increasing the air velocity, which further increases the jet velocity and consequently increases the lift force. The model results simulate very well with CFD results. In Fig. 13, the coefficient of drag (C_D) is plotted against the jet velocity. It can be observed that C_D and C_L appear specifically variable and dependent on the intensity of the Coanda Jet. The front air intake has also the advantage of allowing an effective reduction of Drag by means of the front air intake.

The results show clearly the expected lift and drag results. It can be observed clearly the concomitant presence of the three key phenomena that characterize the system: suction by the jet source, Coanda adhesion, and attraction of the surrounding fluid. Considering C_L, it can be determined the value of C_L at different values of the speed of the jet. In this case, it can be observed an effective increase of the lift coefficient on the basis of the results reported in Fig. 14.

The obtained results for the lift and drag force on the reference wing are given in Fig. 15. It can be said that the model simulates very well with CFD study.

It can be expressed the aerodynamic drag and lift into two different ways: One hypothesis is to assume the original wing without any propulsion as a reference; the second is to assume the dual wing without any jet.

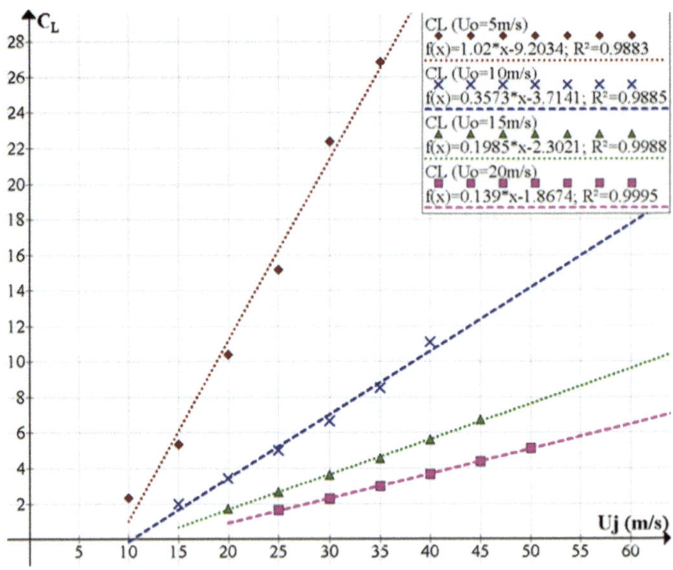

Fig. 14 C_L as function of jet speed at different velocities

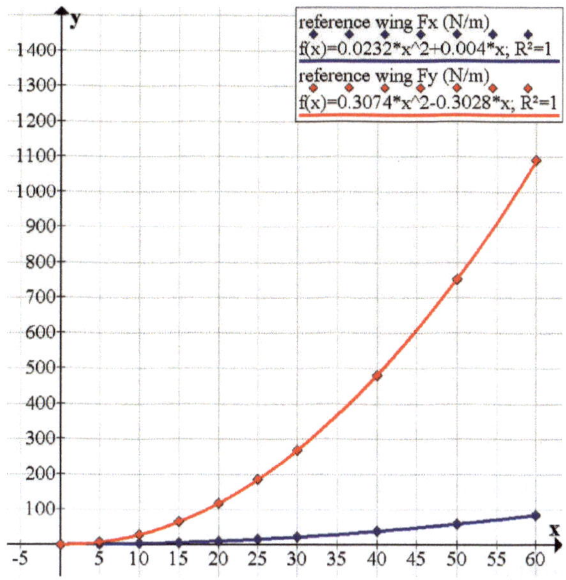

Fig. 15 Drag and lift computed on the reference wing

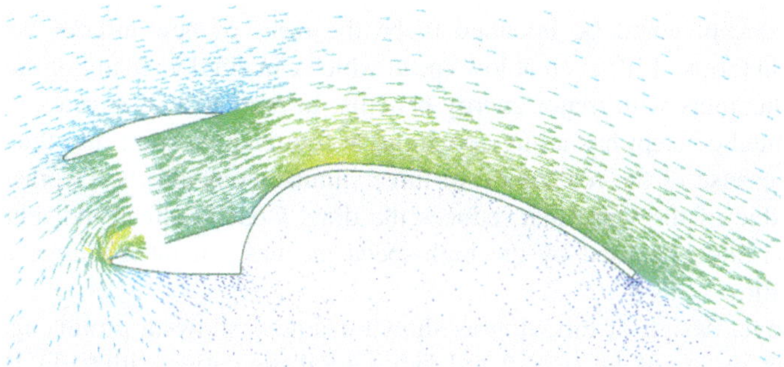

Fig. 16 Velocity vectors in the flow domain

Fig. 16 shows the velocity vectors, higher velocity and adhesion of the jet are found at the Coanda surface. The gauge pressure contours (Fig. 17) also depict this fact.

Fig. 17 Pressure contours in the flow field

7 Conclusions

This chapter has presented a new propelled wing concept with a breakthrough potential in terms of lift generation, much higher than any other wing concept ever designed before.

This concept could be intended to be the basis of new aircraft with a huge potential in terms of lift even at low speed which can emulate many of the operation of the helicopters with major energy benefits. In particular, it appears evident that the presented concept has a huge potential of future application into the aeronautic domain because of the realization of three simultaneous effects: Depression at the leading edge of the wing that reduces the drag, Coanda effect and related super-circulation effects caused by the high-speed jet that generates a huge increase in terms of lift.

The novel design of the wing as shown in Fig. 7 shows a promising candidate for the new design of the aircraft and shows good simulation with CFD. However, it will require further validation through 3D CFD study and experiments, which are on the way to be performed.

Acknowledgements The authors are highly indebted to the Sheffield Hallam University for providing the ANSYS Fluent 17.1 and to Prof. Antonio Gameiro Lopez of University of Coimbra for the possibility of using EasyCFD.

References

1. McGowen SS (2005) Helicopters: an illustrated history of their impact. ABC-CLIO, Inc.
2. Munson K (1973) Helicopter and other rotorcrafts since 1907. Blandford Press, London

3. Raptis IA, Valavanis KP (2011) Linear and nonlinear control of small-scale unmanned helicopters. International series on intelligent systems, control and automation: science and engineering, vol 45. Springer, Berlin
4. Mill ML et al (1967) Helicopters calculation and design, vol. I aerodynamics. NASA technical translation, NASA TT F-494
5. Norton B (2004) Bell/Boeing V-22 osprey: multi service tiltrotor, 1st edn. Aerofax Series, Midland
6. Thornborough AM (1990) Bell-Boeing V-22 osprey: an aeroguide special, 1st edn. Linewrights Ltd.
7. Gertler J (2011) V-22 osprey Tilt-rotor aircraft: background and issues for congress. CRS Report for Congress, Congressional Research Service 7-5700, 6 Oct 2011
8. Trancossi M (2011) An overview of scientific and technical literature on coanda effect applied to nozzles. SAE technical papers no. 2011-01-2591. ISSN 0148-7191
9. Trancossi M, Dumas A (2011) ACHEON: Aerial coanda high efficiency orienting-jet nozzle. SAE technical papers no. 2011-01-2737
10. Abdollahzadeh M et al (2015) Numerical design and analysis of a multi-DBD actuator configuration for the experimental testing of ACHEON nozzle model. Aerosp Sci Technol 41:259–273
11. Páscoa JC, Brójo FMP, Monteiro JM M (2009) Numerical simulation of magneto-plasma thrusters for aerospace propulsion using and mhd formulation paper O-7.2. In: Proceedings 14th international conference on emerging nuclear energy systems, Instituto Tecnológico e Nuclear
12. Trancossi M, Stewart J, Maharshi S, Angeli D (2016) Mathematical model of a constructal Coanda effect nozzle. J Appl Fluid Mech 9(6):2813–2822
13. Trancossi M (2015) Design of ACHEON thrust and vector propulsion system. SAE Technical paper no. 2015-01-2425
14. Trancossi M et al (2015) Preliminary implementation study of ACHEON thrust and vector electrical propulsion on a STOL light utility aircraft. SAE Technical paper no. 2015-01-2422
15. Trancossi M et al (2016) A new aircraft architecture based on the ACHEON Coanda effect nozzle: flight model and energy evaluation. Eur Transp Res Rev 8(2):1–21
16. Trancossi M, Stewart J, Pascoa JC (2016) A new propelled wing aircraft configuration. International mechanical engineering congress and exposition. American Society of Mechanical Engineers, pp V001T03A048–V001T03A048
17. Radespiel R, Burnazzi M, Casper M, Scholz P (2016) Active flow control for high lift with steady blowing. Aeronaut J 120(1223):171–200
18. Landa T, Radespiel R, Wild J (2016) Numerical simulations of stream wise vortices on a generic high-lift configuration. 54th AIAA Aerospace sciences meeting 2016:0304
19. Stoll AM (2015) Comparison of CFD and experimental results of the LEAPTech distributed electric propulsion blown wing. In: 15th AIAA aviation technology, integration, and operations conference, p 3188
20. Clarke S, Papathakis K, Samuel A, Lin Y, Ginn S (2016) NASA SCEPTOR electric concept aircraft power system: X-plane electric propulsion system design and qualification for crewed flight-testing. In: Transportation electrification conference and expo (ITEC), pp 1–27
21. Drăgan V (2013) A new mathematical model for high thickness Coanda effect wall jets. Rev Air Force Acad
22. Roderick WEB (1961) Use of the Coanda effect for the deflection of jet sheets of a smoothly curve surfaces, part II. University of Toronto, Institute of Aerophysics, Technical Note no.5
23. Benner SD (1965) The Coandă effect at deflection surfaces widely separated from the jet nozzle, University of Toronto. UTISA Technical Note no. 78
24. Drăgan V (2012) Contributions regarding the design of a self super circulated rotary wing. In: Mod Tech international conference
25. Schlichting H (1979) Boundary layer theory, 7th edn. McGraw-Hill, New York
26. VV.AA (2016) Ansys Fluent 17.1 User Guide, Ansys Inc., USA

27. Lopes AMG (2010) A versatile software tool for the numerical simulation of fluid flow and heat transfer in simple geometries. Comput Appl Eng Educ 18(1):14–27
28. Casey M, Wintergerste T (2000) ERCOFTAC special interest group on quality and trust in industrial cfd: best practice guidelines, Version 1.0
29. Rizzi A, Vos J (1998) Towards establishing credibility in computational fluid dynamics. AIAA J 36(5):668–675
30. Celik I, Li J, Hu G, Shaffer C (2005) Limitations of Richardson Extrapolation and Some Possible Remedies. J Fluids Eng 127:795–805
31. Spalart P, Allmaras S (1992) A one-equation turbulence model for aerodynamic flows. 30th Aerospace sciences meeting and exhibit: 439
32. Menter FR, Kuntz M, Langtry R (2003) Ten years of industrial experience with the SST turbulence model. Turbul Heat Mass Transfer 4(1):625–632
33. Menter FR (1994) Two-equation eddy-viscosity turbulence models for engineering applications. AIAA J 32(8):1598–1605
34. Versteeg HK, Malalasekaera W (2007) An introduction to computational fluid dynamics the finite volume method, 2nd edn. Pearson Education Limited
35. Patankar SV (1980) Numerical heat transfer and fluid flow. Hemisphere, Washington, DC

Stochastic Modeling in Industry and Management

Aliakbar Montazer Haghighi and Dimitar P. Mishev

1 Basics of Probability

A result of performing an experiment is called an *outcome* of that experiment. If results are not deterministic, the experiment is called a *random experiment*. Some particular sets of outcomes of an experiment are considered *simple* if they will not be able to break into smaller sets. Such sets are called *simple events*. Often, simple events are just singletons (a set with a single element, such as $\{s\}$). An *event* is a combination of simple events. By an *occurrence of an event,* it is meant appearance of a result of the random experiment. The set of all possible outcomes of a random experiment is called the *sample space*. Thus, an event is a subset of the sample space. An element of the sample space is referred to as a *sample point*. Two events are called *mutually exclusive* if their intersection is the empty set.

By definition, the *probability of an event* is a number between 0 and 1 (inclusive). It describes the likelihood of occurrence of the event. Consider a sample space with finite number of elements, say n. If all outcomes of such a sample space have the same probability to occur, then a probability of $1/n$ is assigned to each point and the sample space is called *equiprobable*. Hence, the probability of an event with k points will be k/n. The expression "*at random*" is also used for equiprobable sample space. This definition can be extended to an infinite sample space. In that case, elements are referred to as with *uniform measure*. In summary, the *probability of an event* meant the *ratio of the number of ways the event can occur to the total number of ways any outcome can occur*.

A.M. Haghighi (✉) · D.P. Mishev
Department of Mathematics, Prairie View A&M University, Prairie View, TX, USA
e-mail: amhaghighi@pvamu.edu

D.P. Mishev
e-mail: dimichev@pvamu.edu

M. Ram and J.P. Davim (eds.), *Modeling and Simulation in Industrial Engineering*,
Management and Industrial Engineering, DOI 10.1007/978-3-319-60432-9_7

The theory of probability and consequently stochastic processes (that will be defined later) is based on a set of three fundamental axioms, originated by the Russian mathematician Kolmogorov [1]. These axioms are defined on the triple $(\Omega, \mathfrak{B}, P)$, called the *probability space*, where Ω is the sample space, $\mathfrak{B}$ is the set function containing all possible events drawn from Ω, and P is the probability of an event. The first axiom states that probability of an event from the sample space is a number between 0 and 1, exclusive. The second says that the probability of the sample space is 1. Finally, the third axiom states that probability of two mutually exclusive events from the sample space is the sum of probabilities of each of the two events.

If A and B are two events from the sample space $\mathfrak{B}$, the probability of occurrence of both of these events, say $P(A)$ and $P(B)$, at the same time is denoted by $P(AB)$. The probability of B given that the event A has already occurred is denoted by $P(B|A)$, called the *conditional probability* of B given A and is defined as:

Interval [0, 1]

$$P(B|A) = \frac{P(AB)}{P(A)}, \quad \text{for any event } A \text{ in } \mathfrak{B}, \tag{1.1}$$

provided that $P(A) > 0$. If $P(A) = 0$, then $P(B|A)$ is not defined. It is obvious that $P(A|A) = 1$. Given conditional probability as the measure, the sample space Ω reduces to $(\Omega, \mathfrak{B}, P(\cdot|A))$. We say that two events A and B are *independent* if and only if $P(AB) = P(A)P(B)$. The latter elation can be expanded for an arbitrary number of events. Also, for two events A and B, we have $P(A|B) = P(A)$. Thus, from (1.1) we have $P(AB) = P(B|A)P(A) = P(A|B)P(B)$. An important property in the theory of probability is *The Law of Total Probability* or *Total Probability Theorem* that states if $A_1, A_2, \ldots, A_n$ are n mutually exclusive events whose sum is 1, then for any given arbitrary event A we have

$$P(A) = \sum_{i=1}^{n} P(A_i)P(A|A_i). \tag{1.2}$$

Relation (1.1) can be extended for more than two events. For instance, for three events A, B, and C, the probability of occurrences of all three events at the same time will be:

$$P\{A \text{ and } B \text{ and } C\} = P\{A\}P\{B|A\}P\{C|A \text{ and } B\}. \tag{1.3}$$

Example 1.1 Suppose that a company has posted an ad for several managerial positions. An applicant has gone through a chain of interviews and has predicted that his chance of getting the job is 85%. During the interviews, the applicant was told that 40% of hired mangers will be given temporary housing until they find their homes. However, of those receiving housing 30% will have to share their temporary houses. The applicant is to calculate his/her chance of getting a job, temporary housing, and no house sharing.

To calculate the probability in question, let the events of being accepted, getting housing, and no sharing (that is, individually) be denoted by A, H, and I, respectively. Thus, from (1.3) we will have:

$$\begin{aligned} P\{A \text{ and } B \text{ and } C\} &= P\{A\}P\{B|A\}P\{C|A \text{ and } B\} \\ &= (0.85)(0.40)(1-0.30) = 0.238 = 2.38\%. \end{aligned}$$

In other words, this applicant has about 2% chance to get a job, get housing and live by himself/herself.

2 Discrete Random Variables and Discrete Probability Distribution Functions

A random variable is defined as a function, say X, that assigns to a sample point a numerical value (or a set of values). If X is defined on an at most countable sample space into the set of real numbers, it is called a *discrete random variable*. By the probability distribution of a discrete random variable X, it is meant assignment of probabilities over the entire values of X. The *probability mass function (pmf) of* X, denoted by p_x, $p_x > 0$ is defined as $p_x = P(X = x)$. Of course sum of all p_x over x is 1. When two random variables X and Y have the same distribution, say p_X and p_Y, we say X and Y are *equally distributed.*

We are to discuss four important discrete distributions that will be now used in this chapter. They are Bernoulli, binomial, geometric, and Poisson.

A *Bernoulli trial* (or experiment) is a trial with exactly two possible outcomes. The outcomes are referred to as "*success*" with probability p, and "*failure*" with probability $1 - p$. Thus, if X is a random variable taking values 1 and 0, corresponding to success and failure, respectively, with $p(p > 0)$ as probability of success and $1 - p$ as probability of failure, then, we will have:

$$P(X = k) = p^k(1-p)^{1-k}, \quad k = 0, 1. \tag{2.1}$$

Formula (2.1) is the *probability distribution function (pmf) of the Bernoulli random variable X.*

If a Bernoulli trial is repeated independently n times with the same probabilities of success and failure on each trial, then, the process is called *Bernoulli trials.*

We define the *arithmetic average* for k integers $n_1, n_2, \ldots, n_k$, denoted by $\bar{n}$, as the sum of the integers divided by k. That is,

$$\bar{n} = n_1 \cdot \frac{1}{k} + n_2 \cdot \frac{1}{k} + \cdots + n_k \cdot \frac{1}{k}. \tag{2.2}$$

Similarly, if X a discrete random variable with values $x_1, x_2, \ldots, x_k$ with probabilities $p_1, p_2, \ldots, p_k$, then, the *mean* (or *weighted average* or *expected value* or *expectation* or *mathematical expectation*) of the discrete random variable X, denoted by $E(X)$, is defined as

$$E(X) = \sum_{i=1}^{k} x_i p_i. \tag{2.3}$$

Relation (2.3) was defined for a discrete random variables of finite values. We can extend (2.3) for a discrete random variable with infinite values $x_0, x_1, x_2, \ldots$, and respective probabilities $p_0, p_1, p_2 \ldots$; that is, $P(X = x_i) = p_i, i \in \mathbb{N}_0$ For such a case, the mean of X is defined as

$$E(X) = \sum_{i=0}^{\infty} x_i p_i, \tag{2.4}$$

provided that the infinite series in (2.4) converges absolutely, otherwise, the series does not exist and, hence, the mean of X does not exist.

Some basic properties of mean are as follows:

i. Let X and Y be discrete random variables and a, b, and c constants, then

$$E(aX + bY + c) = aE(X) + bE(Y) + c. \tag{2.5}$$

ii. Let X and Y be two independent discrete random variables with each individual pmf, that is referred to as *marginal pmf*, p_X and p_Y, respectively. The pair (X, Y) is a random variable with values from the domain of each of the variable. Distribution of (X, Y) is referred to as the *joint distribution*, denoted by $p_{X,Y}$.

Let us also assume that $E(X)$ and $E(Y)$ both exist. Then,

$$E(XY) = E(X)E(Y). \tag{2.6}$$

iii. For discrete random variables X and Y, with x and y as their values, respectively, we have

$$E(X|Y = y) = \sum_{x} xP(X = x|Y = y), \tag{2.7}$$

Example 2.1 Let X and Y be two discrete random variables with values $X = 1, 2, 3$, and $Y = 1, 2, 3, 4$. Let also the joint distribution of X and Y be given in Table 1.

Table 1 Joint probability distribution and marginal pmf

Y	X			
	1	2	3	p_Y
1	0.20	0.05	0.10	0.35
2	0.03	0.01	0.15	0.19
3	0.07	0.02	0.20	0.29
4	0.03	0.04	0.10	0.17
p_X	0.33	0.12	0.55	1

From Table 1, we can find conditional probabilities. For instance, to calculate $p_{x|3}$, we first use the definition of conditional probability (1.1) for the random variables, take the value of p_3 from the last column of Table 1, and find the value of each $p_{x,3}/p_3$ for each x from Table 1, row 3 and each value of the corresponding column. Thus, we can have:

$$p_{x|3} = \frac{p_{x,3}}{p_Y} = \frac{p_{x,3}}{0.29} = \begin{cases} \frac{0.07}{0.29} = 0.24, & x = 1, \\ \frac{0.02}{0.29} = 0.07, & x = 2, \\ \frac{0.20}{0.29} = 0.69, & x = 3, \\ 0, & \text{otherwise} \end{cases}.$$

We can also calculate conditional expected value, say $E(Y|X = 3)$, as follows:

$$E(Y|x = 3) = \sum_y y p_{y|3} = (1)(0.10) + (2)(0.15) + (3)(0.20) + (4)(0.1) - 1.4.$$

3 Moment and Probability Generating Function

Relations (2.3) and (2.4) can be expanded for higher powers of the discrete random variable X. The expected value of X, $E(X)$, is sometimes referred to as the first moment of X. Expected values of higher powers of X are similarly named with the powers of X. Thus,

$$E(X), E(X^2), E(X^3), \ldots, E(X^n), \tag{3.1}$$

are called, respectively, *the first moment of X, the second moment of X, the third moment of X*, …, and *the* n*th moment of X*.

Let us now assume that $\mu = E(X)$. The nth moment of the random variable $X - \mu$, that is, $E[(X - \mu)^n]$, is called the *central moment* of X. The random variable $X - \mu$ measures the *deviation of X from its mean*.

The *mean square deviation*, that is, $E\left[(X - \mu)^2\right]$, is called the *second central moment of* X. It is also known as the *variance* of X. Thus, the variance of a random variable measures the *average deviation* or *dispersion* of the random variable from its mean. The positive square root of the variance of a random variable X is called the *standard deviation* of X.

Let X be a discrete random variable with nonnegative integer values with pmf of X as p_X. The *probability generating function* of X, denoted by $G(z)$, where z is a complex number is defined as

$$G(z) \equiv E(z^X) = \sum_{x=0}^{\infty} p_X z^x, \tag{3.2}$$

where the power series in (3.2) converges absolutely at least for all z such that $|z| \leq 1$.

Note that $G(1) = 1$ because sum of probabilities is 1.

Example 3.1 Let us toss a fair die. By being fair, it is meant that all six sides have equal chance to occur. That is, probability of each side to occur is 1/6. Thus, probability generating function of the random variable representing occurrence of sides of the die will be:

$$\begin{aligned} G(z) &= \sum_{x=0}^{\infty} p_x z^x \\ &= \frac{1}{6}z + \frac{1}{6}z^2 + \frac{1}{6}z^3 + \frac{1}{6}z^4 + \frac{1}{6}z^5 + \frac{1}{6}z^6 + 0z^n, \quad n = 1, 7, 8, \ldots. \end{aligned}$$

Of course, $G(1) = \frac{1}{6} + \frac{1}{6} + \frac{1}{6} + \frac{1}{6} + \frac{1}{6} + \frac{1}{6} = 1$.

4 Selected Standard Discrete Distribution Functions

4.1 Bernoulli Probability Distribution Function

The random variable representing a trial with only two outcomes is called a *Bernoulli random variable,* and such experiment is called a *Bernoulli trial.* The sample space for such a trial has only two sample points (or outcomes). The two sample points are usually referred to as the *success* and the *failure*. The experiment in this case is called a *Bernoulli trial*. If X is a Bernoulli random variable with values 1 and 0, corresponding to success and failure, respectively, with probabilities p, $0 \leq p \leq 1$, and $1 - p$, respectively, then distribution of X is given by:

$$P(X = 1) = p; \quad P(X = 0) = 1 - p, \tag{4.1}$$

or

$$P(X = k) = p^k (1 - p)^{1-k}, \quad k = 0, 1, \tag{4.2}$$

with $p + (1 - p) = 1$.

The relation (4.2) defines a distribution function and is called the *Bernoulli probability distribution function.*

4.2 Binomial Probability Distribution Function

Let us consider a Bernoulli trial represented by a desecrate random variable X with probabilities of success and failure as p, $0<p<1$, and $1-p$, respectively. We represent n independent repetition of X, with the same probabilities in each trial, by X_n. Suppose the number of success desired within the n trials is k. Then, the *pmf* of X_n is denoted by $\mathcal{B}_k \equiv b(k; n, p)$ and is called the *binomial distribution function* **with** *parameters n* **and** *p* of the random variable X. The X in this case is called the *binomial random variable* and sometimes is referred to as $X \sim b(k; n, p)$. It can easily be shown that

$$\mathcal{B}_k \equiv b(k; n, p) = \binom{n}{k} p^k (1-p)^{n-k}, \quad k = 0, 1, 2, \ldots, n, \tag{4.3}$$

with

$$\sum_{k=0}^{n} \binom{n}{k} p^k (1-p)^{n-k} = 1. \tag{4.4}$$

4.3 Geometric Probability Distribution Function

Geometric distribution finds the probability of the first success in a number of Bernoulli trials. Let X be a discrete random variable with *pmf* as:

$$f(k; 1, p) = P(x = k) = p(1-p)^k, \quad k = 0, 1, \ldots. \tag{4.5}$$

Then, *pmf* given by (4.5) is called a *geometric probability distribution function.*

We note that historically, geometric probability was first studied in the eighteenth century. *Buffon's needle problem* that asks what is the probability that a needle hits randomly onto an area marked with equal parallel lines will cross one of the lines?

Example 4.1 Let us denote by $P_k, k = 1, 2, \ldots$ the geometric probability mass function given in (4.5). Choosing $p = 0.4$, the first 10 terms of (4.5) are listed in Table 2. Also, Fig. 1 shows both probability mass function and cumulative probability function for geometric distribution. Since probabilities are 0 after $k = 18$, we have chosen infinity as 20.

Table 2 Probabilities of the number of applications in FS, Example 1.4.1

P_1	P_2	P_3	P_4	P_5	P_6	P_7	P_8	P_9	P_{10}
0.4000	0.2400	0.1440	0.0864	0.0518	0.0311	0.0187	0.0112	0.0067	0.0040

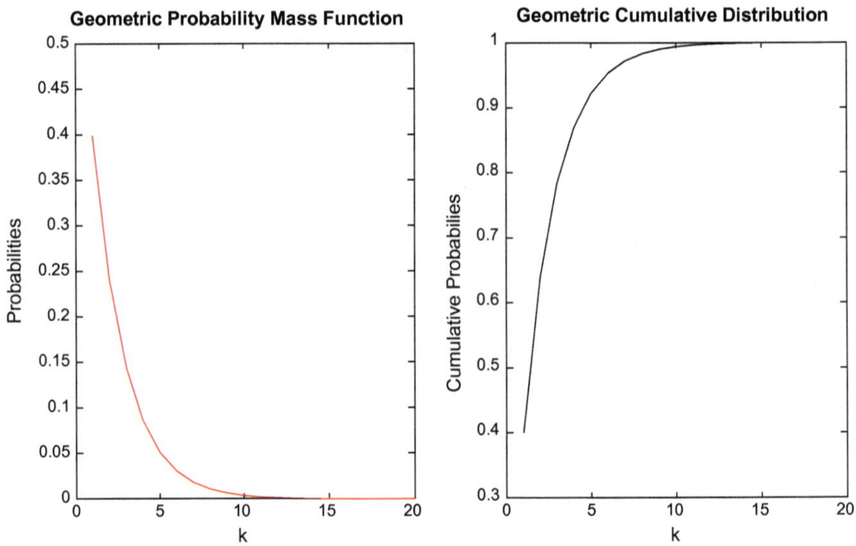

Fig. 1 Geometric probability mass function

4.4 Poisson Probability Distribution Function

A discrete random variable X with pmf as

$$p_x = P(X = k) = \frac{e^{-\lambda}\lambda^k}{k!}, \quad k = 0, 1, 2, \ldots, \tag{4.6}$$

where λ is a constant, is a called *Poisson random variable,* and (4.6) is called *Poisson distribution function with parameter* λ.

We note that Poisson distribution finds the probability of a given number of events such as arrivals, occurring in a given interval of time or space. Historically, this distribution was introduced in 1837 by a French mathematician Siméon Poisson.

French Mathematician, Siméon Denis Poisson (1781–1840) **From:** https://en.wikipedia.org/wiki/Sim%C3%A9on_Denis_Poisson

Example 4.2 Suppose that orders for a product to a manufacture arrives at a receiver 5 times per hour, on the average. A manager is looking to calculate the probability that in a given hour 6 orders of the product will be recorded.

To help the manager, let X be a random variable representing the arrival of orders, then denoting the probability in question by p_6, that is, $p_6 = P\{X = 6\}$, from (4.6), this probability can be calculated as follows:

$$p_6 = P\{X = 6\} = \frac{e^{-5}5^6}{6!} = 0.1462.$$

5 Continuous Random Variables and Continuous Probability Distribution Functions

Now, suppose that the values of outcomes of a random experiment are real numbers. Then, the sample space, Ω, is a referred to a *continuous sample space*. Since, the set of all subsets of real numbers is infinitely large, assigning probabilities to all points, the set that contains all events of interest, called Borel sets, will be chosen. Let *f*(*x*) be defined on the set of real numbers, $\mathbb{R}$, such that $f(x) \geq 0$, for all real *x*, and $\int_{-\infty}^{\infty} f(x)\mathrm{d}x = 1$. Then, *f*(*x*) is called a *continuous probability density function* (*pdf*) on $\mathbb{R}$ and it is denoted by $f_X(x)$. The random variable *X* having this property is referred to as a *continuous random variable*. In this case, we write:

$$P(a \leq X \leq b) \int_a^b f_X(x)\mathrm{d}x, \quad \text{for any interval [a,b]}, \tag{5.1}$$

Accordingly, the *probability distribution function* or the *cumulative distribution function* (*cdf*) of X, denoted by $F_X(x)$, is defined as

$$F_X(x) = P(X \le x) = \int_{-\infty}^{x} f(t)\mathrm{d}t. \tag{5.2}$$

The following are two properties of (5.2):

(i) $$P(a \le X \le b) = P(a < X \le b) = P(a \le X < b) = P(a < X < b). \tag{5.3}$$

(ii) If $a \le b$, then

$$P(a \le X \le b) = P(X \le b) = P(X \le a) = F_X(b) - F_X(a). \tag{5.4}$$

(iii) It can easily be verified that

$$F_X(-\infty) = 0 \quad \text{and} \quad F_X(\infty) = 0. \tag{5.5}$$

6 Selected Standard Continuous Distribution Functions

6.1 Uniform Probability Distribution Function

A continuous random variable X with the following probability density function:

$$f_X(x) = \begin{cases} \frac{1}{b-a}, & a \le x \le b, \\ 0 & \text{otherwise.} \end{cases} \tag{6.1}$$

over an interval $[a,\ b]$ is called the *uniform distribution random variable*. Integrating (6.1) yields the *uniform distribution function of* X as:

$$F_X(x) = \begin{cases} 0, & x \le a, \\ \frac{x-a}{b-a}, & a \le x \le b, \\ 1, & x \ge b. \end{cases} \tag{6.2}$$

6.2 (Negative) Exponential Probability Distribution Function

A continuous random variable X with *pdf*

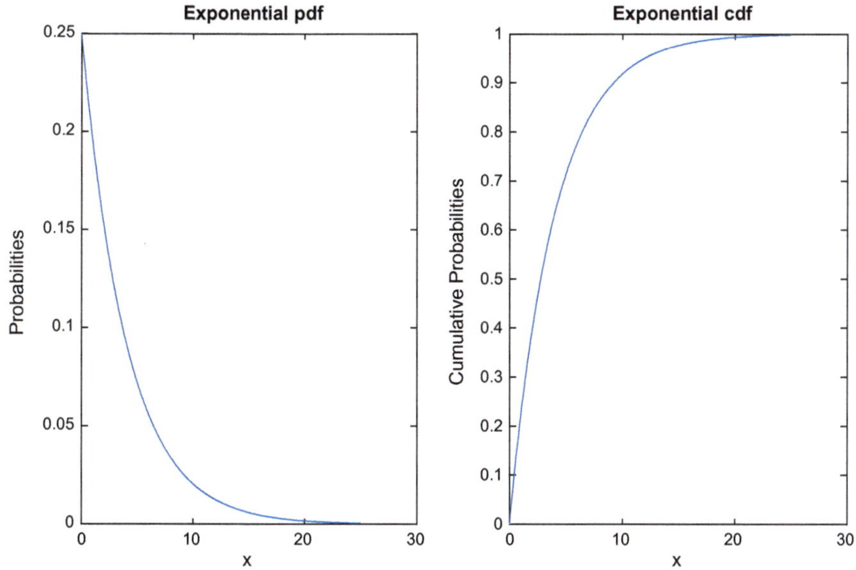

Fig. 2 Exponential distribution function with $\mu = 4$

$$f_X(t) = \begin{cases} \mu e^{-\mu t}, & t \leq 0, \\ 0, & \text{otherwise.} \end{cases} \tag{6.3}$$

and *cdf*

$$F_X(t) = \begin{cases} 1 - e^{-\mu t}, & t \leq 0, \\ 0, & \text{otherwise.} \end{cases} \tag{6.4}$$

is called *negative exponential* (or *exponential*) *random variable* with parameter μ. It can easily be seen that $1/\mu$ is the mean of this random variable.

Example 6.1 Assuming $\mu = 4$, pdf and cdf for exponential distribution function are given in Fig. 2.

Relation (6.4) is the distribution function of the random variable that is a special case of the random variable called *gamma random variable* with pdf

$$f_X(x; \mu.t) = \begin{cases} \frac{\mu^t x^{t-1}}{\Gamma(t)}, & x \geq 0, \\ 0, & \text{otherwise,} \end{cases} \tag{6.5}$$

and cdf:

$$F_X(x; \mu.t) = \begin{cases} \frac{1}{\Gamma(t)} \int\limits_0^x \mu^t u^{t-1} e^{-\mu u} du, & x \geq 0, \\ 0, & \text{otherwise,} \end{cases} \tag{6.6}$$

is called a *gamma random variable with parameters* μ *and t,* with $\mu > 0$ and $t \geq 1$, where $\Gamma(t)$ is defined by:

$$\Gamma(t) = \int\limits_0^\infty x^{t-1} e^{-x} dx, \tag{6.7}$$

where t is a positive real number.

Example 6.2 Let us return to Example 4.2 and suppose that time to fill an order is 10 min, on the average. The manager is interested to calculate the probability that an agent fills an order by less than 7 min.

Once again to help the manager, let T be a random variable representing the time of filling an order. Then, denote the probability in question by F_7, that is, $F_7 = P\{X < 7\}$. It is clear that the rate of filling orders is 1/10. Thus, from (6.4), the probability in question can be calculated as follows:

$$F(7) = 1 - e^{-(0.1)(7)} = 0.5034.$$

7 Stochastic Processes

A *stochastic process* is a sequence or a family of random variables, say $\{X(t), t \in T\}$, where t is an element of an index set (or the *parameter space*) T (denumerable or nondenumerable). The index t could be an epoch of time. The process is called *discrete-time* or *continuous-time* depending upon the index set as it is discrete or continuous, respectively. When the index set, T, is nonnegative, the discrete-time process is denoted by $\{X_n, n = 0, 1, \ldots\}$. On the other hand, for the continuous-time, it is denoted by $\{X(t), t \geq 0\}$. The set of all possible values of random variable $X(t)$, denoted by S, is called the *state space* of the stochastic process. S may be finite, countable, or noncountable.

A discrete-time stochastic process $X_n, n = 0, 1, \ldots\}$ with a state space $S = \{x_0.x_1, \ldots\}$ is called *discrete-time Markov chain* or simply a *Markov chain*, if we are following the property, called the *Markov Property* which holds:

$$P\{X_n = x_n | X_0 = x_0, X_1 = x_1, \ldots, X_{n-1}| = x_{n-1}\} = P\{X_n = x_n | X_{n-1}| = x_{n-1}\}. \tag{7.1}$$

The Markov property (7.1) states that *given the current state of the process, transition to the next state independent of the history of the process prior to the current state*. This property is called the "*forgetfulness*" property of the Markov chain.

Russian Mathematician, Andrey (Andrei) Andreyevich Markov (Markoff): June 14, 1856–July 20, 1922 From: https://en.wikipedia.org/wiki/Andrey_Markov#/media/File:AAMarkov.jpg

We note that the exponential distribution function is the only continuous probability distribution having the forgetfulness property and conversely a continuous probability distribution function having such a property is exponential. Also, for the discrete case, only the geometric mass function has the forgetful property.

The right-hand side of (7.1) is called a *one-step transition probability* from state x_{n-1} to state x_n. In general; the one-step transition probability from state x to state y, denoted by p_{xy}, is defined as:

$$P_{xy} = P\{X_n = y | X_{n-1} = x\}, \quad x, y \in S. \tag{7.2}$$

If the relation (7.2) is independent of time, the Markov chain is called *homogeneous* and is denoted by p_{xy}. In other words, the Markov chain is homogeneous if:

$$P_{xy} = P\{X_{n+j} = y | X_{n+j-1} = x\}, \quad x, y \in S, \quad j = 0, \pm 1, \ldots, \pm(n-1), \ldots. \tag{7.3}$$

The *n-step* transition probability from state x to state y, denoted by $p_{xy}^{(n)}$, is defined as:

$$P_{xy}^{(n)} = P\{X_{j+n} = y | X_j = x\}, \quad x, y \in S, \quad n = 0, 1, \ldots, \tag{7.4}$$

where by convention we have

$$p_{xy}^{(0)} = \begin{cases} 1, & y = x, \\ 0, & y \neq x. \end{cases} \tag{7.5}$$

There are similar properties for continuous-time process as we stated for discrete-time. The stochastic process of continuous-time is denoted by $\{X(t), t \geq 0\}$ and is called a *Markov process*. The *transition probability* from x to y in a time interval of length t will be denoted by $p_{xy}(t)$, and it means that $p_{xy}(t) = P\{X(s+1) = y|X(s) = x\}$. Similar definitions and formulas for Markov chain follow accordingly by replacing n by t. For instance, the continuous case of (7.1) is:

$$P\{T > t+s|T > t\} = P\{T > s\}, \quad \forall s, t \geq 0. \tag{7.6}$$

Transition probabilities may be represented by matrix, referred to as the *transition matrix*, denoted by $\mathbf{P}(t) = (p_{xy})$, where $p_{xy}(t)$ is the general element. We note that transition probabilities are nonnegative and sum to 1. It is assumed that at least one transition from state x to another state in [0, t], for all $t \geq 0$ occurs. For this reason, the transition matrix **P**(t) is called a *stochastic matrix*. The transition probabilities satisfy what is called the *Chapman-Kolmogorov equations*

$$p_{xy}(t+t) = \sum_{k=0}^{\infty} p_{xk}(t) p_{ky}(s) \tag{7.6}$$

or in matrix form

$$\mathbf{P}(t+s) = \mathbf{P}(t)\mathbf{P}(t), \quad \forall s, t \in [0, \infty]. \tag{7.7}$$

Similar equations can be derived for discrete-time Markov chains.

A Markov process that is independent of time is referred to as the *stationary* or *time-homogeneous* or simply *homogeneous*.

Example 7.1. Maze as a Markov Chain A standard example used for a Markov chain is "*Rat in a Maze.*" We use it as a puzzle or a game that a *player* can used for fun, yet with substantial mathematics involved.

Suppose we have a maze consisting of 5 cells, say, Cell 1, Cell 2, …, Cell 5, with an exit through Cell 5 that we call it Cell 0. See Fig. 3a for a three-dimensional and Fig. 3b for a two-dimensional maze described in this example. Figure 4 is a symbolic picture of rats in mazes. It is assumed that the player cannot remember in what cell he was after a move (transition) has occurred (the Markov property). We also assume that his next move is independent of his previous moves. We further assume that choices to move out of a cell from each pathway out to neighboring cells are equally likely.

Let us denote the transition probabilities from Cell i to Cell j by $p_{i,j}, i,j = 0, 1, 2, \ldots, 5$. The probability $p_{0,0}$ will stand for an exit which will be permanent. Thus, $p_{0,0} = 1$ Since no transition is possible from state 0, this state is referred to as the *absorbing state*. In other words, as soon as the chain enters this state, it will

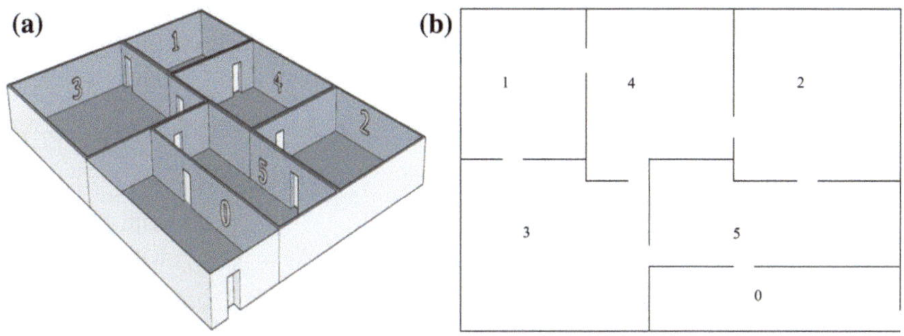

Fig. 3 **a** Graph of a three-dimensional maze with 5 cells, Example 7.1. **b** Graph of a two-dimensional maze with 5 cells, Example 7.1

Fig. 4 Two simple three-dimensional maze pictures found on the Internet from the link: https://www.google.com/search?site=&source=hp&q=rat+in+a+maze&oq=rat+in+a+maze&gs_l=hp.3…3447.8779.0.9329.14.14.0.0.0.0.327.565.2-1j1.2.0….0…1.1.64.hp..12.1.326.0..0j0i20k1j0i131k1.xqnv9b98-Dc

remain there forever. In terms of (7.6), we denote by X_j, the cell that player will visit immediately after the jth move. Thus, the process $\{X_j\}$ is a Markov chain with state space $S = \{0, 1, 2, 3, 4, 5\}$. We assume that staying put or moving from a cell to itself in the next move is not allowed, that is $p_{i,i} = 0, i = 1, 2, \ldots, 5$.

The structure of the maze is as follows (numbers and j within the brackets indicate cell numbers):

$$\text{Cell}\,0 \rightarrow \begin{cases} \text{Cell}\,0, & \text{with } p_{0,0} = 1, \\ \text{Cell}\,j, & \text{with } p_{0,j} = 0,\ j = 1, 2, \ldots, 5. \end{cases} \quad \text{or} \quad X_0 = 0,\ \text{with } p_{0,0} = 1. \tag{7.8}$$

$$\text{Cell}\,1 \rightarrow \begin{cases} \text{Cell}\,3, & \text{with } p_{1,3} = 1/2, \\ \text{Cell}\,4 & \text{with } p_{1,4} = 1/2, \\ \text{Cell}\,j, & \text{with } p_{1,j} = 0,\ j = 0, 1, 2, 5. \end{cases} \quad \text{or} \quad X_1 \begin{cases} 3, & \text{with } p_{1,3} = 1/2, \\ 4, & \text{with } p_{1,4} = 1/2. \end{cases} \tag{7.9}$$

$$\text{Cell } 2 \rightarrow \begin{cases} \text{Cell } 4, & \text{with } p_{2,4} = 1/2, \\ \text{Cell } 5, & \text{with } p_{2,5} = 1/2, \\ \text{Cell } j, & \text{with } p_{2,j} = 0, \ j = 0,1,2,3. \end{cases} \quad \text{or} \quad X_2 \begin{cases} 4, & \text{with } p_{2,4} = 1/2, \\ 5, & \text{with } p_{2,5} = 1/32 \end{cases} \tag{7.10}$$

$$\text{Cell } 3 \rightarrow \begin{cases} \text{Cell } 1, & \text{with } p_{3,1} = 1/3, \\ \text{Cell } 4, & \text{with } p_{3,4} = 1/3, \\ \text{Cell } 5, & \text{with } p_{3,5} = 1/3, \\ \text{Cell } j, & \text{with } p_{3,j} = 0, \ j = 0,2,3. \end{cases} \quad \text{or} \quad X_3 = \begin{cases} 1, & \text{with } p_{3,1} = 1/3, \\ 4, & \text{with } p_{3,4} = 1/3, \\ 5, & \text{with } p_{3,5} = 1/3. \end{cases} \tag{7.11}$$

$$\text{Cell } 4 \rightarrow \begin{cases} \text{Cell } 1, & \text{with } p_{4,1} = 1/3, \\ \text{Cell } 2, & \text{with } p_{4,2} = 1/3, \\ \text{Cell } 3, & \text{with } p_{4,3} = 1/3, \\ \text{Cell } j, & \text{with } p_{4,j} = 0, \ j = 0,4,5. \end{cases} \quad \text{or} \quad X_4 = \begin{cases} 1, & \text{with } p_{4,1} = 1/3, \\ 2, & \text{with } p_{4,2} = 1/3 \\ 3, & \text{with } p_{4,3} = 1/3. \end{cases} \tag{7.12}$$

$$\text{Cell } 5 \rightarrow \begin{cases} \text{Cell } 0, & \text{with } p_{5,0} = 1/3, \\ \text{Cell } 2, & \text{with } p_{5,2} = 1/3, \\ \text{Cell } 3, & \text{with } p_{5,3} = 1/3, \\ \text{Cell } j, & \text{with } p_{5,j} = 0, \ j = 1,4,5. \end{cases} \quad \text{or} \quad X_5 = \begin{cases} 0, & \text{with } p_{5,0} = 1/3, \\ 2, & \text{with } p_{5,2} = 1/3, \\ 3, & \text{with } p_{5,3} = 1/3. \end{cases} \tag{7.13}$$

Thus, the transition probability matrix, denoted by **P**, will be as follows:

$$\mathbf{P} = \begin{array}{c} \\ 0 \\ 1 \\ 2 \\ 3 \\ 4 \\ 5 \end{array} \begin{array}{c} \begin{array}{cccccc} 0 & 1 & 2 & 3 & 4 & 5 \end{array} \\ \begin{pmatrix} 1 & 0 & 0 & 0 & 0 & 0 \\ 0 & 0 & 0 & 1/2 & 1/2 & 0 \\ 0 & 0 & 0 & 0 & 1/2 & 1/2 \\ 0 & 1/3 & 0 & 0 & 1/3 & 1/3 \\ 0 & 1/3 & 1/3 & 1/3 & 0 & 0 \\ 1/3 & 0 & 1/3 & 1/3 & 0 & 0 \end{pmatrix} \end{array} \tag{7.14}$$

Note that matrix (7.14) is different from Table 1, which is the table of joint probabilities. Thus, from matrix (2.1), we can calculate conditional probabilities. For instance, values of $p_{x|5}$ are those in the last row of matrix **P**, that is,

$$p_{x|5} = \begin{cases} 1/3, & x = 0, \\ 0, & x = 1, \\ 1/3, & x = 2, \\ 1/3, & x = 3, \\ 0, & x = 4, \\ 0, & x = 5. \end{cases} \tag{7.15}$$

Now, the question is what is the mean number of moves for the player to exit the maze starting from any of the cells, assuming that he initially started from Cell $i, i = 1,2,3,4,5$?

To answer this question, we formulate the problem as follows: Denoting the minimum number of moves for the player to exit, if he began from Cell $i, i = 1,2,3,4,5$, by $\mu_{i,0}, i = 1,2,3,4,5$, where $\mu_{i,0}, i = 1,2,3,4,5$, is defined by

$$\mu_{i,0} = \min\{j \geq 0 \text{ such that } X_j = 0|X_0 = i, i = 1,2,3,4,5\}, \tag{7.16}$$

calculate the values of $E(\mu_{i,0}), i = 1,2,3,4,5$.

So, let us start with $i = 1$, that is, we assume the player initially stands in Cell 1, i.e., $X_0 = 1$, where he starts moving to the possible neighboring cells. Hence, we will find $E(\mu_{1,0})$. From (7.7) through (7.15), we see that the player cannot directly exit from Cell 1, it rather has to move around cells to get to Cell 5, before he will have a chance to exit. However, we know that from Cell 1, he can move to Cell 3 with one move and probability 1/2. Thus, $E(\mu_{1,0}|X_1 = 3) = 1 + E(\mu_{3,0})$. Since from Cell 1 he could also go to Cell 4 with 1 move and with probability 1/2, we will have $E(\mu_{1,0}|X_1 = 4) = 1 + E(\mu_{4,0})$. Here is the role of the Markov property. That is, once the player is in Cell 3, for example, he will not remember how he got there, that is, in which cell he was before or even before that. Thus, the chance of remaining number of moves to exit from the cell he is in now would be the same as if he were in Cell 1. Based on this argument, we will have the following:

$$\begin{aligned} E(\mu_{1,0}) &= E(\mu_{1,0}|X_1 = 3)P\{X_1 = 3|X_0 = 0\} + E(\mu_{1,0}|X_1 = 4)P\{X_1 = 4|X_0 = 0\} \\ &= \left(\frac{1}{2}\right)(1 + E(\mu_{3,0})) + \left(\frac{1}{2}\right)(1 + E(\mu_{4,0})) \\ &= \left(\frac{1}{2}\right)1 + E(\mu_{3,0}) + \left(\frac{1}{2}\right)E(\mu_{4,0}). \end{aligned} \tag{7.17}$$

In a similar fashion, we can find the rest of the expected values, except the last one, $E(\mu_{5,0})$. Thus, we will have:

$$\begin{aligned} E(\mu_{2,0}) &= E(\mu_{2,0}|X_2 = 4)P\{X_2 = 4|X_0 = 0\} + E(\mu_{2,0}|X_2 = 5)P\{X_2 = 5|X_0 = 0\} \\ &= \left(\frac{1}{2}\right)(1 + E(\mu_{4,0})) + \left(\frac{1}{2}\right)(1 + E(\mu_{5,0})) \\ &= 1 + \left(\frac{1}{2}\right)E(\mu_{4,0}) + \left(\frac{1}{2}\right)E(\mu_{5,0}). \end{aligned} \tag{7.18}$$

$$\begin{aligned} E(\mu_{3,0}) &= E(\mu_{3,0}|X_3 = 1)P\{X_3 = 1|X_0 = 0\} + E(\mu_{3,0}|X_3 = 4)P\{X_3 = 4|X_0 = 0\} \\ &\quad + E(\mu_{3,0}|X_3 = 5)P\{X_3 = 5|X_0 = 0\} \\ &= \left(\frac{1}{3}\right)(1 + E(\mu_{1,0})) + \left(\frac{1}{3}\right)(1 + E(\mu_{4,0})) + \left(\frac{1}{3}\right)(1 + E(\mu_{5,0})) \\ &= 1 + \left(\frac{1}{3}\right)E(\mu_{1,0}) + \left(\frac{1}{3}\right)E(\mu_{4,0}) + \left(\frac{1}{3}\right)E(\mu_{5,0}). \end{aligned} \tag{7.19}$$

$$\begin{aligned} E(\mu_{4,0}) &= E(\mu_{4,0}|X_4 = 1)P\{X_4 = 1|X_0 = 0\} + E(\mu_{4,0}|X_4 = 2)P\{X_4 = 2|X_0 = 0\} \\ &\quad + E(\mu_{4,0}|X_4 = 3)P\{X_4 = 3|X_0 = 0\} \\ &= \left(\frac{1}{3}\right)(1 + E(\mu_{1,0})) + \left(\frac{1}{3}\right)(1 + E(\mu_{2,0})) + \left(\frac{1}{3}\right)(1 + E(\mu_{3,0})) \\ &= \left(\frac{1}{3}\right)1 + E(\mu_{1,0}) + \left(\frac{1}{3}\right)E(\mu_{2,0}) + \left(\frac{1}{3}\right)E(\mu_{3,0}). \end{aligned} \tag{7.20}$$

$$\begin{aligned} E(\mu_{5,0}) &= E(\mu_{5,0}|X_5 = 0)P\{X_5 = 0|X_0 = 0\} + E(\mu_{5,0}|X_5 = 2)P\{X_5 = 2|X_0 = 0\} \\ &\quad + E(\mu_{5,0}|X_5 = 3)P\{X_5 = 3|X_0 = 0\} \\ &= \left(\frac{1}{3}\right)(1) + \left(\frac{1}{3}\right)(1 + E(\mu_{2,0})) + \left(\frac{1}{3}\right)(1 + E(\mu_{3,0})) \\ &= 1 + \left(\frac{1}{3}\right)E(\mu_{2,0}) + \left(\frac{1}{3}\right)E(\mu_{3,0}). \end{aligned} \tag{7.21}$$

Therefore, we have 5 equations with 5 unknowns

$$E(\mu_{1,0}), E(\mu_{2,0}), E(\mu_{3,0}), E(\mu_{4,0}), E(\mu_{5,0}) \tag{7.22}$$

as follows:

$$\begin{cases} E(\mu_{1,0}) - \left(\frac{1}{2}\right)E(\mu_{3,0}) - \left(\frac{1}{2}\right)E(\mu_{4,0}) = 1, \\ E(\mu_{2,0}) - \left(\frac{1}{2}\right)E(\mu_{4,0}) - \left(\frac{1}{2}\right)E(\mu_{5,0}) = 1, \\ -\left(\frac{1}{3}\right)E(\mu_{1,0}) + E(\mu_{3,0}) - \left(\frac{1}{3}\right)E(\mu_{4,0}) - \left(\frac{1}{3}\right)E(\mu_{5,0}) = 1, \\ -\left(\frac{1}{3}\right)E(\mu_{1,0}) - \left(\frac{1}{3}\right)E(\mu_{2,0}) - \left(\frac{1}{3}\right)E(\mu_{3,0}) + E(\mu_{4,0}) = 1, \\ -\left(\frac{1}{3}\right)E(\mu_{2,0}) - \left(\frac{1}{3}\right)E(\mu_{3,0}) + E(\mu_{5,0}) = 1. \end{cases} \tag{7.23}$$

Thus,

$$\begin{cases} E(\mu_{1,0}) = \frac{120}{13} = 8.85, \\ E(\mu_{2,0}) = \frac{115}{13} = 9.23, \\ E(\mu_{3,0}) = \frac{92}{13} = 7.08, \\ E(\mu_{4,0}) = \frac{122}{13} = 9.38, \\ E(\mu_{5,0}) = \frac{82}{13} = 6.31. \end{cases} \tag{7.24}$$

What (7.24) says is that if the player starts in Cell 1, he will exit after about 9 moves, on the average. The same for others, that is, if the player starts in Cell 2, he will exit after about 9 moves, on the average, after about 7 moves if he starts in Cell 3, after about 9 moves if he starts in Cell 4, and after about 6 moves if he starts in Cell 5.

8 Generating Functions and Their Application in Solution of Difference Equations

Consider a sequence of real numbers $\{a_n\}$ with $a_n = 0, n < 0$. The function then $G(z)$ defined by the infinite series

$$G(z) \equiv \sum_{n=0}^{\infty} a_n z^n, \tag{8.1}$$

is called the *generating function* of the sequence $\{a_n\}$, if it converges for $|z| < R$, where R is the radius of convergence.

We can define (8.1) for a discrete random variable X. Let us consider the random variable z^X with *pmf* p_n, where n is a nonnegative integer. Then, the *probability generating function* or (if there is no fear of confusion) simply the *generating function* of X, denoted by $G(z)$, is defined as:

$$G(z) \equiv E(Z^X) = \sum_{n=0}^{\infty} p_n z^n, \tag{8.2}$$

where $E(X) = \sum_{x=0}^{\infty} x p_x$. As it can been seen from (8.2), p_n can be found by expanding it and evaluating it at $z = 0$.

9 Queueing Processes

An example of stochastic process, we will discuss the queueing processes and present several models. We use the word "queue" to mean what has been used over a century, started with the work of A.K. Erlang (1878–1929) that is, "waiting line." See Erlang [2]. The theory of queues deals with the stochastic behavior of processes arising in connection with mass servicing in cases when random fluctuation occurs, such as telephone traffic, particle counting, processes of automatic ticket machines, stock processes.

Danish Mathematician Agner Krarup Erlang: January 1, 1878–February 3, 1929. From: https://en.wikipedia.org/wiki/Agner_Krarup_Erlang

9.1 Mechanism of a Queue

Basic mechanism of a queue consists of *Arrival, service,* and *departure* processes. The *queue discipline* is the order of service. For instance, service may be conducted in order of arrival, i.e., *first-come first-served* (*FCFS*) or *first-in-first-out* (*FIFO*), random service, priority service, *last-come first-served* (*LCFS*), *batch* (*bulk*) *service*, etc. Of course, different types of features may be added to a model such as arrival singly or by bulk, service singly or by batch, service with vacation, parallel or series multiple servers, server and feedback with finite or infinite buffer, feedback, splitting. We note that the words "bulk," "batch," and "group" are used interchangeably. The batch sizes may be fixed or variable. In some cases, the batch size may have a minimum and a maximum.

To each of the processes within a queue model, there may be associated a particular or general distribution. For instance, an arrival process may be according to a Poisson distribution and service may be conducted according to an exponential distribution. Feedback may be according to a Bernoulli or geometric distributions, and splitting may be according to Bernoulli distribution.

10 Certain Stochastic Models in Industry and Management

In this part, we present some stochastic models, each as an application in queueing theory. Thus, in some we will find the distribution of the number of tasks in the system, while in others we will find the distribution of the system busy period as well as the servers' busy period.

10.1 M/M/1 *Queueing Model*

We start briefly discussing the simplest queueing model to see its formulization and solution. It will help understanding the more complicated models. The model is described as follows: customs arrive according to a Poisson distribution with parameter λ. Service in this model is by a single server and serves according to an exponential distribution with parameter μ. Service is based on the first-come and first–served. This simple queue is denoted by *M*/*M*/1. It reads Markovian arrival, Markovian service, and single server. Such a model is an example of a stochastic process called birth and death process. Some of the measures to be found are distribution of the number of customers in the system (queue length), average length of time the server is busy and average queue length.

The goal is to formulate this mode and find distribution of the number of customers in the system at any time; it is assume that over time, we look at the number of jobs in the system, called *state of the system* and possible *transitions* at that point. Thus, if it is desired to have k jobs in the system at the observation time, there are three possibilities: (1) There are already k jobs in the system and, thus, no transaction should occur, that is, no arrival and no departure, (2) the system is short of one job, that is there are $k - 1$ jobs in the system, and thus, one arrival should occur, or (3) the system has one job over, that is there are $k + 1$ jobs in the system and, hence, one service should be completed and depart. Therefore, the system may be looked at the most general type of *birth and death process* (*B-D*) that allows only transitions of the form $k \to k+1$, $k \to k$, or $k \to k-1$, at any time t and travels back and forth among the integers in one-step transitions, if not staying in the same place. Roughly speaking, a birth-death process is a process which, over time, moves among the nonnegative integers or some ordered subset of the nonnegative integers in a one-step transition. Such a process does not allow transitions of more than one step. In other words, the probability of a transition of more than one step is considered to be negligible.

Based on the discussion above and *Chapman-Kolmogorov's backward and forward equations,* the system of differential difference equation for such a system is, then, developed and solved using different methods, including generating function and matrix methods. Formulations of the system for discrete- and continuous-time are very much similar. For detail, see Haghighi and Mishev [3].

To set up the stage for more advances cases, here we show how the system of differential difference equations for *M*/*M*/1 is developed.

We should note that it is easy to prove that for a Poisson arrival process, the interarrival times follow an exponential distribution with the same parameter. Using the exponential distribution, we can describe the state of the system at time t in a simple way, namely the number of customers in the system. Also, we do not have to remember when the last customer arrived or when the last customer entered service. This is because the exponential distribution is memoryless.

Now let us denote by $p_n(t)$ the probability that at time t there are n customers in the system. The system capacity assumed to be infinite implies that $n = 0, 1, 2, \ldots$. The memoryless property of the exponential distribution implies that the system forges the history. Thus, in the related conditional probabilities only the current, one state back, and one state forward would matter. Thus, the system of differential difference equations within a small period of time Δt is as follows:

$$\begin{cases} p_0(t+\Delta t) = \lambda(1 - \Delta t)p_0(t) + \mu\Delta t p_1(t) + o(\Delta t), \\ p_n(t+\Delta t) = \lambda\Delta t p_{n-1}(t) + [1 - (\lambda+\mu)\Delta]t)p_n(t) + \mu\Delta t p_{n+1}(t) + o(\Delta t), n = 1, 2, \ldots, \end{cases} \tag{10.1.1}$$

where $o(\Delta t)$ is called the "*little o*" and is defined as

$$\lim_{\Delta t \to 0} = \frac{o(\Delta t)}{\Delta t} = 0. \tag{10.1.2}$$

Letting $\Delta t \to 0$ in (10.1.1), we obtain the system of differential difference equations for the *M*/*M*/1 queueing system as follows:

$$\begin{cases} P_0'(t) = -\lambda p_0(t) + \mu p_1(t), \\ P_n'(t) = \lambda p_{n-1}(t) - (\lambda+\mu)p_n(t) + \mu p_{n+1}(t), \quad n = 1, 2, \ldots. \end{cases} \tag{10.1.3}$$

If the derivatives on the left hand of (10.1.3) are set to 0, we obtain a set of difference equations called the *stationary or time-independent system of difference equations*, as follows:

$$\begin{cases} \lambda p_0 = \mu p_1, \\ (\lambda + \mu) p_n = \lambda p_{n-1} + \mu p_{n+1}, \quad n = 1, 2, \ldots. \end{cases} \tag{10.1.4}$$

Of course, the sum of probabilities must be zero. This fact is referred to as *the normalizing equation*, which is

$$\sum_{n=0}^{\infty} p_n = 1. \tag{10.1.5}$$

The system (10.1.4) can easily be solved by recursion method. Let us define $\rho = \lambda/\mu$. For the system to have a solution ρ must be less than 1. Now from the first equation of (10.1.4), we will have:

$$p_1 = \rho p_0. \tag{10.1.6}$$

Substituting (10.1.6) in the second equation of (1.10.4), we obtain p_2, and continuing this way, we will have

$$p_n = \rho^n p_0, \quad n = 0, 1, 2, \ldots. \tag{10.1.7}$$

Using (10.1.5), from (10.1.6), we obtain:

$$p_0 + \rho p_0 + \rho^2 p_0 + \cdots = 1 \tag{10.1.8}$$

or

$$p_0 = 1 - \rho. \tag{10.1.9}$$

Thus,

$$p_n = (1 - \rho)\rho^n, \quad n = 0, 1, 2, \ldots. \tag{10.1.10}$$

There are other methods such as generating function method to solve the system (10.1.4). To see this method, denoting the probability generating function for p_n by $G(z)$, $|z| < 1$, we define $G(z)$ as:

$$G(z) = \sum_{n=0}^{\infty} p_n z^n, \quad |z| < 1. \tag{10.1.11}$$

Applying (10.1.11) on (10.1.4), after some algebra, we obtain:

$$G(z) = \sum_{n=0}^{\infty} (1 - \rho)\rho^n z^n = \frac{1 - \rho}{1 - \rho z}. \tag{10.1.12}$$

Thus, by writing the McLaurin expansion of (10.1.12), we can see the values of p_0 and p_n that we found in (10.1.9) and (10.1.10).

Derivatives of the generating function (10.1.12) lead us to find the all moments. For instance, the mean and variance of number of the jobs in the system, that is, in the waiting line and in the service. Thus, denoting the number of jobs in the system by N, the mean by L_s, and variance by $\text{Var}_s(N)$, we will have:

$$L_s = E(N) = \frac{\rho}{1-\rho} = \frac{\lambda}{\mu - \lambda} \quad \text{and} \quad V_s = \text{Var}_s(N) = \frac{\rho}{(1-\rho)^2}. \tag{10.1.13}$$

We could also find the average and the variance of length of the jobs waiting in line, q, denoted by L_q and V_q, respectively, as:

$$L_q = L_q + \rho. \tag{10.1.14}$$

Hence,

$$L_q = L_s - \rho = \frac{\lambda^2}{\mu(\mu - \lambda)} \quad \text{and} \quad V_q = \frac{\rho^2(1+\rho-\rho^2)}{(1-\rho)^2}. \tag{10.1.15}$$

To find the average time spent in the system, denoted by W_s,, there is a very important formula, called *Little's formula*, that gives the relation between the average number of customers in the system and the average time spent in the system. It is:

$$L_s = \lambda W_s. \tag{10.1.16}$$

Thus, from (10.1.13) and (10.1.16) we can find the average of time a job will be in the system, as:

$$W_s = \frac{L_s}{\lambda} = \frac{1}{\mu - \lambda}. \tag{10.1.17}$$

Since the average service time is $1/\mu$, we can also find the average waiting time in the waiting line, denoted by W_q, as well as the variance, denoted V_q, as follows:

$$W_s = W_q + \frac{1}{\mu} \quad \text{and} \quad W_q = W_s - \frac{1}{\mu} = \frac{\lambda}{\mu(\mu - \lambda)}. \tag{10.1.18}$$

The variance of length of the time a job spends in the system, denoted by $VW_s = \text{Var}(W_s)$, and the variance of length of the time for the jobs waiting in line, denoted by $VW_q = \text{Var}(W_q)$, respectively, are:

$$VW_s = \text{Var}(W_s(N)) = \frac{1}{(1-\rho)^2\mu^2} \quad \text{and} \quad VW_q = \frac{\rho(2-\rho)}{(1-\rho)^2\mu^2}. \tag{10.1.19}$$

We note that in case the buffer is finite, say with capacity *N*, then the notation *M*/*M*/1/*N* is used and the system (10.1.4) can be adjusted accordingly and we will have the distribution of the number of jobs in the system, the mean number of jobs in the system, and the mean number of jobs lost due to the finiteness of the buffer, respectively, as:

$$p_{Nn} = \begin{cases} \frac{1-\rho}{1-\rho^{N+1}}\rho^n, & \text{if } \rho \neq 1, \quad n = 0, 1, 2, N, \\ \frac{1}{N+1}, & \text{if } \rho = 1, \quad n = 0, 1, 2, N. \end{cases} \tag{10.1.20}$$

$$L_{N_s} = \begin{cases} \frac{\rho}{1-\rho} - \frac{(N+1)\rho^{N+1}}{1-\rho^{N+1}}, & \rho \neq 1, \\ \frac{N}{2}, & \rho = 1. \end{cases} \tag{10.1.21}$$

$$L_{N\,\text{loss}} = \begin{cases} \frac{\lambda(1-\rho)\rho^N}{1-\rho^{N+1}}, & \text{if } \rho \neq 1, \\ \frac{\lambda}{N+1}, & \text{if } \rho = 1. \end{cases} \tag{10.1.22}$$

Note that as $N \to \infty$, (10.1.20) approaches (10.1.10) and so they both would be the same, as expected.

Example 10.1.1 Let us consider an example for the system *M*/*M*/1 and *M*/*M*/1/*N*. Suppose the average arrival rate $\lambda = 20$ jobs per unit time and the average service rate $\mu = 40$ jobs per unit time. Hence, $\rho = \lambda/\mu = 1/2 < 1$. Thus, from the obtained information and using MATLAB 2017a, with $N = 20$ (as infinity) for *M*/*M*/1 and two cases for *M*/*M*/1/*N* one with $N = 4$ and another with 20 (as infinity), we will have the measures, Table 3 and Fig. 5.

M/*M*/1:

$$L_s = 1, \quad V_s = 2, \quad L_q = 0.5, \quad V_q = 1.25, \quad W_s = 0.05,$$
$$W_q = 0.025, \quad VW_s = 0.0025, \quad VW_q = 0.0019.$$

M/*M*/1/*N*:

M/*M*/1/4: $L_{N_s}s = 0.8387$, $L_{N\,\text{loss}}s = 0.6452$. *M*/*M*/1/20: $L_{N_s}s = 1$, $L_{N\,\text{loss}}s = 0.0000095367$.

10.2 A Multiserver Queueing System with Balking and Reneging

About six decades ago, the idea of a customer not joining a waiting line due to the length of the line or other reasons as well as changing his/her mind once in the line

Table 3 Probability distribution for $M/M/1$ and $M/M/1/N$ and when N approaches infinity

$M/M/1$ $n = 0, 1, 2, \ldots, 19$	p_n	$M/M/1/N$, $N = 4$ $n = 0, 1, 2, 3, 4$ P_{Nn}	$M/M/1/N$, $N = 20$ $n = 0, 1, 2, \ldots, 19$ P_{Nn}
0	0.5000	0.5161	0.5000
1	0.2500	0.2581	0.2500
2	0.1250	0.1290	0.1250
3	0.0625	0.0645	0.0625
4	0.0313	0.0323	0.0313
5	0.0156		0.0156
6	0.0078		0.0078
7	0.0039		0.0039
8	0.0020		0.0020
9	0.0010		0.0010
10	0.0005		0.0005
11	0.0002		0.0002
12	0.0001		0.0001
13	0.0001		0.0001
14	0.0000		0.0000
15	0.0000		0.0000
16	0.0000		0.0000
17	0.0000		0.0000
18	0.0000		0.0000
19	0.0000		0.0000
Sum	1.0000	1.0000	1.0000

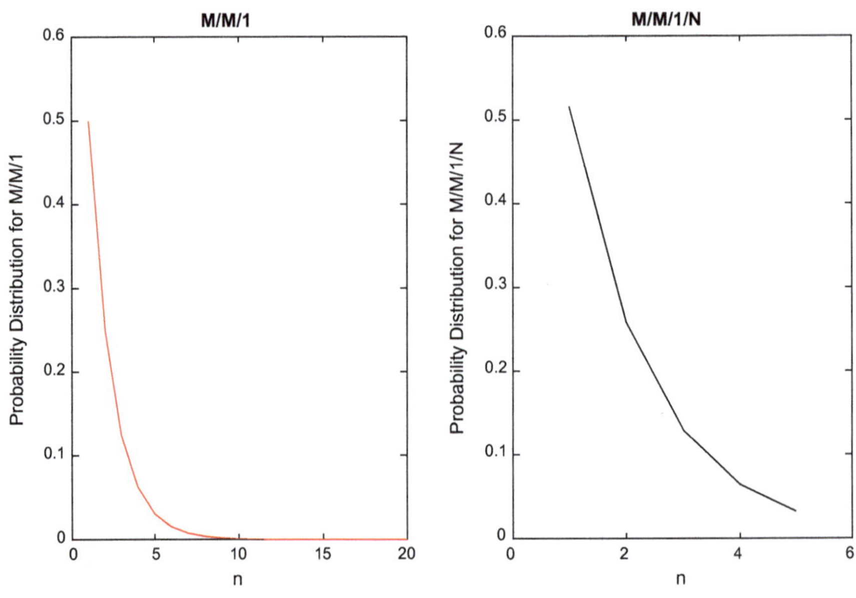

Fig. 5 Probability distribution for $M/M/1$ and $M/M/1/N$

was studied. For instance, see Haight [4]. This is a very practical problem. A very simple example of this case is when you call a large company, say ATT, who has many answering services who are set in parallel all over the world. When you call, you will be told how long your waiting time will be before someone will talk with you or how many other customers are ahead of you. Thus, based on the waiting time or the number ahead of you, you may be discouraged and hang up, or you may stay online anyway. But, after a while, you change your mind and hang up. Haight [4, 5] considers a model such that each arriving customer will join the waiting line if its maximum length is K and not join, if the queue length exceeds K. He assumed K to be a random variable.

The idea was developed through the years and many considered different scenarios of the case. See for instance, Ancker Jr and Gafarian [6], and Montazer-Haghighi [7]. Here, we offer the case Montazer-Haghighi et al. [8] and Haghighi and Mishev [3, 9] considered.

Before we do that, however, we want to define two terms. They are balking and reneging. We use the word *balking* for discouragement before joining a waiting line (queue) and *reneging* when a customer joins, waits a while, then leaves before getting any service.

10.2.1 The Model

Jobs arrive from an infinite source according to a Poisson distribution of parameter λ. There are m servers set in parallel. An arriving job may balk with a constant probability b when all servers are busy at its arrival and will join the system with probability p at such point, $b + p = 1$. Hence, attending rate of jobs is $p\lambda$. Of course, the instantaneous balking rate would be $b\lambda$. It is assumed that the waiting room or the *buffer* is of infinite capacity.

After joining the system, a job will wait in the waiting line for a while, assuming all servers are busy, and will leave the system before receiving service. The length of stay before leaving for a job is assumed to be a random variable that follows an exponential distribution with parameter β.

Service at each service station is also assumed to be exponential with parameter μ. Jobs are served based on the first-come first-served (FCFS). Thus, we may refer to this model as an *M/M/m with balking and reneging*.

Finally, it is assumed that a job balks, reneges, and returns, and it will be considered as a new arrival.

10.2.2 Analysis

Let us consider the system in state E_n, that is, we look at the system when there are n in it, meaning in buffer and being served. We are also considering a transition from E_n to E_n. Based on the description of the model, we can see that

$$\text{Average arrival rate} = \begin{cases} \lambda, & 0 \le n \le m-1, \\ p\lambda, & n \ge m. \end{cases} \tag{10.2.1}$$

and

$$\text{Average service rate} = \begin{cases} m\mu, & 0 \le n \le m-1, \\ m\mu + ((n-m)\beta, & n \ge m. \end{cases} \tag{10.2.2}$$

Let us denote by $p_n(t)$ the *transient-state probability* that there are n jobs in the system at time t, where

$$p_n = \lim_{t\to\infty} p_n(t). \tag{10.2.3}$$

Then, the time-independent or *stationary* system corresponding to the transient case for $m \ge 3$ can be obtained as follows (cases $m = 1$ and $m = 2$ can be deal with separately):

$$\begin{cases} \lambda p_0 = \mu p_1, & \\ (\lambda + n\mu)p_n = \lambda p_{n-1} + (n+1)\mu p_{n+1}, & 1 \le n \le m, \\ (p\lambda + m\mu)p_m = \lambda p_{m-1} + (m\mu + \beta)p_{m+1}, & \\ [p\lambda + m\mu + (n-m)\beta]p_n = p\lambda p_{n-1} +]m\mu + (n+1-m)\beta p_{n+1}, & n > m, \end{cases} \tag{10.2.4}$$

with *the normalizing equation*,

$$\sum_{n=0}^{\infty} p_n = 1. \tag{10.2.5}$$

We note that the intensity factor in this case will be

$$\rho = \frac{\lambda}{m\mu}. \tag{10.2.6}$$

To solve the system (10.2.4), from the first two equations of the system we have:

$$p_n = \frac{1}{n!}\left(\frac{\lambda}{\mu}\right)^n p_0, \quad n \le m. \tag{10.2.7}$$

We compute p_{m-1} and p_m from (10.2.7), and substituting in the third equation of the system (10.2.4), we will have:

$$p_{m+1} = \frac{1}{m!}\left(\frac{\lambda}{\mu}\right)^m \left(\frac{p\lambda}{m\mu + \beta}\right) p_0. \tag{10.2.8}$$

Using (10.2.8) and putting $n = m + 1$ in the last equation of (10.2.4), we obtain:

$$p_{m+2} = \frac{(p\lambda)^2}{(m\mu + \beta)(m\mu + 2\beta)} p_m = \frac{1}{m!}\left(\frac{\lambda}{\mu}\right)^m \frac{(p\lambda)^2}{(m\mu + 2\beta)} p_0. \tag{10.2.9}$$

Finally, we will have a general case for $n > m$ as

$$p_n = \frac{1}{m!}\left(\frac{\lambda}{\mu}\right)^m \frac{(p\lambda)^{n-m}}{\prod_{j=1}^{n-m}(m\mu + j\beta)} p_0, \tag{10.2.10}$$

with p_0 calculated from (10.2.5) as:

$$p_0 \left[\sum_{j=0}^{m} \frac{1}{j!}\left(\frac{\lambda}{\mu}\right)^j + \frac{1}{m!}\left(\frac{\lambda}{\mu}\right)^m \sum_{j=1}^{\infty} \frac{(p\lambda)^j}{\prod_{r=1}^{j}(m\mu + r\beta)}\right]^{-1}. \tag{10.2.11}$$

In a special case if we set $m = 1, p = 1$, and $\beta = 0$, from (10.2.10) and (10.2.11) we obtain the distribution for the *M*/*M*/1 (10.1.11). Also, if $p = 1$ and $\beta = 0$, then

$$p_n = \begin{cases} \frac{1}{n!}\left(\frac{\lambda}{\mu}\right)^n p_0, & n \le m, \\ \frac{1}{m! m^{n-m}}\left(\frac{\lambda}{\mu}\right)^n p_0, & n > m \end{cases} \tag{10.2.12}$$

and

$$p_0 = \left[\sum_{n=0}^{m} \frac{1}{n!}\left(\frac{\lambda}{\mu}\right)^n + \frac{1}{m!\left(m - \frac{\lambda}{\mu}\right)}\left(\frac{\lambda}{\mu}\right)^{m+1}\right]^{-1}. \tag{10.2.13}$$

Example 10.2.1 In the previous example, we found the probability distributions and other measures for finite and infinite single-server Markovian queues. We now want to see an example of a multisever queue. Thus, we consider an *M*/*M*/*m* queue without balking and reneging as described in the model above.

Suppose the average arrival rate $\lambda = 5$ jobs per unit time and the average service rate $\mu = 2$ jobs per unit time and there are 4 screening, the applications from the servers sitting in parallel. Hence, $\rho = \lambda/(4\mu) = 0.625 < 1$. Thus, from the obtained information and using MATLAB 2017a, with $N = 20$ (as infinity) for *M*/*M*/4, we will have the following Table 4, measures, and Fig. 6.

Table 4 Probabilities of the number of applications in FS, Example 10.2.1

P_0	P_1	P_2	P_3	P_4	P_5	P_6	P_7	P_8	P_9	P_{10}
0.0737	0.1842	0.1842	0.2303	0.1919	0.1199	0.0750	0.0469	0.0293	0.0183	0.0114

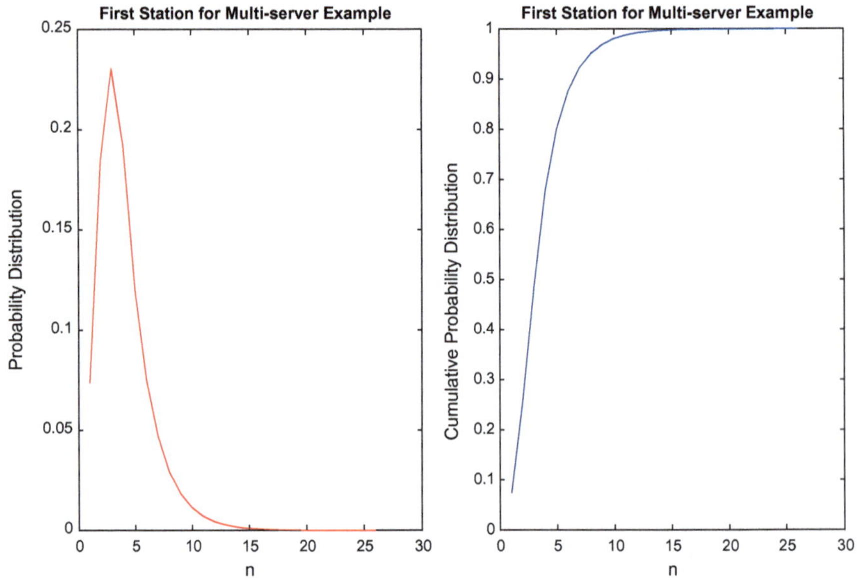

Fig. 6 Graphs of probabilities and cumulative distributions of the number of applications in FS, Example 10.2.1

10.3 A Stochastic Hiring Model

We now present a stochastic hiring process as a network queueing model. In fact, in the current era of technologically oriented communication and manufacturing systems, the performance evaluation is possible by modeling them as queueing networks. Recently, Haghighi and Mishev [10] explored the stochastic hiring process that appeared in part in Haghighi and Mishev [3]. In this section, we highlight the theory in these references and present some numerical example.

10.3.1 The Model

Personnel hiring policy at a large company is considered. It is assumed that there is a constant need for hiring new employees. In other words, hiring process is a continuous task. Applicants apply online through a third party's Web site, an *Inbox*. Generally, many applications arrive at the same time, that is, with bulks of various sizes, during an interval of time, say 24 h. Although in reality, the Inbox's capacity may be limited and, thus, as soon as it is full, the arriving applications are rejected, we design our model with infinite capacity Inbox to avoid loss of application due to capacity.

Now, suppose that there are three stages in the hiring process that are set in tandem. The first stage (referred to as the *FS*) is the initial automated screening process of applications from the Inbox for the applicants' minimum qualifications advertised. A more generalized case can be seen in Haghighi and Mishev [11]. For the sake of analysis, it is assumed that arriving batches of applications to the Inbox follows a time-homogeneous Poisson process with parameter λ. We also assume that processing of a batch in FS follows an exponential distribution with parameter υ. Batches are processed in order of their arrivals; however, there is no ordering process within a batch, that is, processing of applications is by batches, too.

Applications that successfully passed the FS will enter into the second stage (referred to as the **SS**) that consists of a series of interviews. There is also an infinite size inbox, called *buffer*, in SS to hold the arriving batches from the FS. Unsuccessful applications at SF will be discarded. Application batches arrived in the buffer are processed according to the order of their arrivals from FS. Each applicant related to an application in the buffer will go through a number of interviews in sequence before leaving the SS.

The third stage (referred to as the **TS**) is designated to complete the hiring process of the selected candidates. After completion of interviews at SS, applications of applicants who successfully passed the SS will be accepted to be hired and will be sent to the last box referred to as the *H-Box*. Since, practically, the number of available positions are limited, H-Box would be of finite capacity. Thus, in that case the H-Box is full, arriving applications from the SS will be discarded.

10.3.2 The Analysis of the First Stage (FS), Initial Processing

In this subsection, we will briefly analyze the FS and refer to Haghighi and Mishev [10] for full in-depth analysis.

Let us denote the size of an arriving batch by S. When a batch arrives, all applications get to a line in random order and wait to be processed. For the initial screening, the applications from the Inbox are processed with a minimum of b and maximum of $B(B \geq b \geq 1)$ applications. This type of processing is referred to as the *general bulk processing (GBP) rule*, initiated by Neuts [12].

with probability distribution function

$$P\{S = s\} = \alpha_s, \quad s = 1, 2, \ldots \tag{10.3.1}$$

with mean random size of a batch denoted by $\bar{\alpha}$, where

$$\bar{\alpha} = \sum_{s=1}^{\infty} s\alpha_s. \tag{10.3.2}$$

Thus, the arrival process in this case will be a compound Poisson process with arrival rate $\lambda\alpha_s, 0 \leq \alpha_s \leq 1, \sum_{s=1}^{\infty} \alpha_s = 1$, and mean arrival rate $\lambda\bar{\alpha}$. It is assumed that both mean and variance of X are positive and finite. For the initial screening, the applications from the Inbox are processed with a minimum of b and maximum of

B ($B \geq b \geq 1$) applications. This type of processing is referred to as the *general bulk processing (GBP) rule*, initiated by Neuts [12].

The model described, denoted by $M^X/M^{(b,B)}/1$, is a queueing system with the utilization factor

$$\rho = \frac{\lambda\bar{\alpha}}{B\nu} \tag{10.3.3}$$

The goal is to find the probability distribution of the number of applications in FS at any time. Thus, our discussion is to reach this goal. Let the minimum and maximum of batch sizes be b and B, respectively. Now, suppose at the completion of processing of a batch, k, $b \leq k \leq B$, applications will remain in the Inbox. There are three possibilities with k as follows: (1) $0 \leq k \leq b$. Then, the next processing will not start until the number of applications reaches the minimum number b. (2) $b \leq k \leq B$. Then, the processor will take k applications and starts the processing. (3) $k \geq B$. In this case, the processor will take B applications and starts processing them. The rest of applications will remain in the Inbox for the next round of processing.

Let us suppose that we want the probability of j applications in FS at any time, that is, time-independent. Thus, we denote this probability at time t by $P_j(t)$ and its time limit by P_j, $P_j = \lim_{t\to\infty} P_j(t)$. In other words, we are considering the stationary process.

To write the system of balance equations for the system in FS, we have to balance the flow of applications. Thus, we have the following cases: (1) There are i applications in the FS, and we want to empty it. Hence, there is only one possibility and that is the batches of sizes b up to B be processed with rate ν (2) We want to attain n applications in the FS. Hence, either the FS needs i more applications or it has i extra applications. Thus, in the prior case there should be arrivals of applications and in the latter case processing of applications should be completed. Thus, from Haghighi and Mishev [10] the system of balance difference equations for this stationary process that leads to the time-independent distribution of the number of applications in the FS is as follows:

$$\begin{cases} \lambda P_0 = \nu \sum\limits_{j=b}^{B} P_j, & \\ \lambda P_j = \lambda \sum\limits_{i=1}^{j} \alpha_i P_{j-1} + \nu \sum\limits_{i=b}^{B} P_{j+i,}, & \text{if } 1 \leq j \leq b-1, \\ (\lambda+\nu) P_j = \lambda \sum\limits_{i=1}^{j} \alpha_i P_{j-1} + \nu \sum\limits_{i=b}^{B} P_{j+i}, & \text{if } j \geq b. \end{cases} \tag{10.3.4}$$

To solve system (10.3.4), probability generating function (pdf) method is used. Hence, we define the pdfs of α_s and P_j, respectively, as follows:

$$A(z) = \sum_{s=1}^{\infty} \alpha_s z^s \quad \text{and} \quad G(z) = \sum_{j=0}^{\infty} P_j z^j, \quad |z| < 1, \tag{10.3.5}$$

with

$$A(1) = 1,\, A'(1) = \bar{\alpha} \quad \text{and} \quad G(1) = 1. \tag{10.3.6}$$

Hence, after some algebraic manipulations we will have:

$$G(z) = \frac{\nu z^B \sum_{n=0}^{b-1} P_n z^n - \nu \sum_{j=b}^{B} z^{B-j} \sum_{i=0}^{j-1} P_i z^i}{[\lambda - \lambda A(z) + \nu] z^B - \nu \sum_{i=b}^{B} z^{B-i}}, \tag{10.3.7}$$

which coincides with formula (23) of Madan [13]. For the pgf in (10.3.7) to exist, based on the well-known Rouche's theorem, knowing $A(z)$, the number of zeros of denominator within and on the unit circle must be the same as the ones of the numerator. Consequently, the probabilities $P_j, j = 1, 2, \ldots, B - 1$, would be found. Thus, all other unknown probabilities can be found from (10.3.7).

Example 10.3.1 The pgf given in (10.3.7) may become very complicated when one is trying to obtain the probability distribution function. Below we offer a MATLAB computer program using MATLAB version 2017a to find $G(z)$ for a given a set of data. From $G(z)$, probabilities may be calculated by the steps given after the program.

Thus, the goal is to find $P_n, n = 1, 2, 3, \ldots, \infty$. The infinity may be chosen as a number, say *Inf*. Inf may be large enough to make the sum of P_n equal to 1, that is, $\sum_{n=0}^{\infty} P_n = 1$. Here are the steps:

Step 1 **Choose a distribution for the batch sizes**

Suppose that applications arrive in batches of size 2 or 3 according to Bernoulli distribution such that:

$$\begin{cases} \alpha_2 = p, \alpha_3 = 1 - p, & 0 < p < 1, \\ \alpha_s = 0, \text{for} \quad s = 1 & \text{and} \quad s \geq 3. \end{cases} \tag{10.3.8}$$

Hence,

$$\bar{\alpha} = 2p - 3(1 - p) = 3 - p. \tag{10.3.9}$$

Step 2 **Choose a set of data**

We will use the following data:

$$b = 3; \quad B = 5 \quad \lambda = 2, \quad \nu = 7, \quad \text{and} \quad Inf = 20.$$

Hence, from (10.3.3), we should have

$$\rho = \frac{(3-p)\lambda}{v} < 1. \tag{10.3.10}$$

Step 3 **Formulate the generating function of the batch size distribution**
From (10.3.5), we will have the following:

$$A(z) = \alpha_2 z^2 + \alpha_3 z^3 = pz^2(1-z) + z^3. \tag{10.3.11}$$

Step 4 **Write *G*(*z*), symbolically, in MATLAB (Version 2017a)**

```
syms A(z)

p = sym(0.2);
A(z) = p*z^2 + (1-p)*z^3; % The case Bernoulli batch sizes of 2 or 3.

Pn = sym('Pn%d',[1 b]);    % Create vector of unknown constants based on
                           % how many terms are required in the sum

Pi = sym('Pi%d',[1 B]);    % and again

% NUMERATOR FIRST TERM
n = 0:b-1;
numeratorSum1 = nu*z^B*sum(Pn.*(z.^n)); % That is: νz^B ∑_{n=0}^{b-1} P_n z^n

% NUMERATOR SECOND TERM
sum2 = 0;
for j=b:B                           % Outside sum
    i = 0:j-1;
    tmp = sum(Pi(i+1).*(z.^i));     % Inside sum; change i to i+1 due to
                                    % indexing starting from 1
    sum2 = sum2 + z^(B-j).*tmp;     % Term for outside sum
end
numeratorSum2 = -nu*sum2;           % -ν ∑_{j=b}^{B} z^{B-j} ∑_{i=0}^{j-1} P_i z^i

% DENOMINATOR TERM 1
denTerm1 = (lambda-lambda*A(z)+nu)*z^B;   % [λ - λA(z) + ν]z^B

% DENOMINATOR TERM 2
i = b:B;
denTerm2 = -nu*sum(z.^(B-i));             % -ν ∑_{i=b}^{B} z^{B-i}

% CALCULATE G(z)
num = numeratorSum1 + numeratorSum2;      % νz^B ∑_{n=0}^{b-1} P_n z^n - ν ∑_{j=b}^{B} z^{B-j} ∑_{i=0}^{j-1} P_i z^i
den = denTerm1 + denTerm2;                % [λ - λA(z) + ν]z^B - ν ∑_{i=b}^{B} z^{B-i}.

G(z) = num/den;                           % (1.10.3.7)
```

Step 5 Finding P_n using the following steps

(i) Set $P_0 = G(0)$.
(ii) Find the nth derivatives, evaluate at $z = 0$, and divide by $n!$.
(iii) Let $P_n = \frac{G^{(n)}(0)}{n!}$, $n = 1, 2, \ldots, Inf$.

Step 6 **Implementation of steps in Item 5 above**

(i) Take the first derivative, that is, $G'(z)$,
(ii) Substitute 0 for z, that is, find $G'(0)$,
(iii) Divide the result in (ii) by 1!, that is, find $\frac{G'(0)}{1!}$,
(iv) Set $P_1 = \frac{G'(0)}{1!}$.
(v) Repeat items (i) through (iv) in Item 6 above for the second derivative. Thar is, take the second derivative, $G''(z)$,
(vi) Substitute 0 for z, that is, find $G''(0)$,
(vii) Divide the result in v(ii) by 2!, that is, find $\frac{G''(0)}{2!}$,
(viii) Set $P_2 = \frac{G''(0)}{2!}$.
(ix) Repeat (i)–(iv) for the 20th derivative. That is,
(x) Take the 20th derivative, that is, $G^{(20)}(z)$,
(xi) Substitute 0 for z, that is, find $G^{(20)}(0)$,
(xii) Divide the result in (xii) by 20!, that is, find $\frac{G^{(20)}(0)}{20!}$,
(xiii) Set $P_{20} = \frac{G^{(20)}(0)}{20!}$.

Step 7.
Test the result
Sum all $P_n, n = 0, 1, 2, \ldots, Inf$. Result should be 1. If not, the size of Inf may have to be increased.

Example 10.3.2 We let $b = B = 1$. This means that there is no restriction to start processing of batches. That is the FS becomes bulk arrival with single service by a single server. For this case, denoting the mean batch sizes, arrival rate, service rate, utilization factor, pgf for batch size distribution, and pgf for the distribution of the number of application, by $\bar{\alpha}$, λ_1, v_1, ρ_1, $G_1(z)$, and $A_1(z)$, respectively, from (10.3.7) we will have:

$$G(z) = \frac{v_1(1-z)P_0}{v(1-z) - \lambda z[1 - A(z)]}, \tag{10.3.12}$$

which is the same as in Medhi [14]. This is one way for verification of the result obtained.

Now, applying (10.3.6) and L'Hôspital's rule on (10.3.12), we will have

$$\lim_{z\to 1} G(z) = \left.\frac{-\nu P_0}{-\nu - \lambda[1 - A(z) - zA'(z)]}\right|_{z=1} = \frac{-\nu P_0}{-\nu + \lambda A'(1)} = \frac{-\nu P_0}{-\nu + \lambda\bar{\alpha}} = 1. \tag{10.3.13}$$

Hence, from (10.3.13) and (10.3.3), we will have

$$P_0 = 1 - \frac{\lambda\bar{\alpha}}{\nu} = 1 - \rho. \tag{10.3.14}$$

Relation (10.3.14) is the probability of the FS being empty and, thus, of the server being idle. Substituting (10.3.14) into (10.3.12), we can obtain the pgf as follows:

$$G(z) = \frac{1 - \rho}{1 - \frac{\lambda z}{\nu(1-z)}[1 - A(z)]}. \tag{10.3.15}$$

Let us denote the ith moment of the batch size S by $\alpha^{(i)}, i = 1, 2, \ldots$, that is, $E(S^i) = \alpha^{(i)} = 1, 2, \ldots$. Of course, when $i = 1$, $E(S) = \bar{\alpha}$. Now, if we denote the number of applications in the FS by N_1, then we can find the mean and variance of N_1 as:

$$E(N) = \frac{\rho}{1 - \rho}\left(\frac{\alpha^{(2)} + \bar{\alpha}}{2\bar{\alpha}}\right), \tag{10.3.16}$$

and

$$\mathrm{Var}(N) = \frac{\rho}{1 - \rho}\left(\frac{2\alpha^{(3)} + 3\alpha^{(2)} + \bar{\alpha}}{6\bar{\alpha}}\right) + [E(N)]^2, \tag{10.3.17}$$

respectively.

Example 10.3.3 Let us refer to Example 10.3.2. We now consider the case Haghighi and Mishev [3, 10] considered. In this case, it is assumed that batch size distribution is Bernoulli given by

$$\begin{cases} \alpha_1 = p, & \alpha_2 = 1 - p, \quad 0<p<1, \\ \alpha_s = 0, & s \geq 3. \end{cases} \tag{10.3.18}$$

Hence, from (10.3.18), we will have

$$\bar{\alpha} = 1 \cdot p + 2 \cdot (1 - p) = 2 - p. \tag{10.3.19}$$

Also, for this particular case, we will have

$$A(z) = pz + (1-p)z^2, \tag{10.3.20}$$

Thus, from (10.3.15), the pgf is as the follows:

$$G(z) = \frac{1-\rho}{1-\frac{\lambda}{\nu}z - \frac{\lambda(1-p)}{\nu}z^2}. \tag{10.3.21}$$

From (10.3.21), the distribution function of the number of applications in FS, denoted by P_j, is as follows:

$$P_j = \frac{(1-\rho)}{2}\sqrt{\frac{\rho}{\rho+4(2-p)(1-p)}}\left(\frac{\rho}{4-2p}\right)^j \times\left[\left(\sqrt{\frac{\rho+4(2-p)(1-p)}{\rho}}+1\right)^{j+1} + (-1)^j\left(\sqrt{\frac{\rho+4(2-p)(1-p)}{\rho}}-1\right)^{j+1}\right], \quad j = 0, 1, 2, \ldots, \tag{10.3.22}$$

where ρ_1 is given as $\rho = \frac{\lambda\bar{\alpha}}{\nu}$. Hence, from (10.3.16) and (10.3.17), the mean and variance of the number of applications in the FS, in this case (batch size 1 or 2), can be calculated, respectively, as

$$E(N) = \frac{(3-2p)\rho_1}{(2-p)(1-\rho)} \tag{10.3.23}$$

and

$$\mathrm{Var}(N) = \frac{\rho}{(2-p)(1-\rho)}\left[5 - 4p + \frac{(3-2p)^2\rho}{(2-p)(1-\rho)}\right]. \tag{10.3.24}$$

Thus, numerically, if $N = 0, 1\ldots, 25$, $p = 0.2$, $\lambda = 2$, and $\nu = 7$, then $\bar{\alpha} = 2 - p = 1.8$ and $\rho = \frac{\lambda\bar{\alpha}}{\nu} = \frac{18}{35}$. Hence, the probabilities of having 0 up to 25 applications in the FS are listed in Table 5. Additionally, the mean and variance of the number of applications are $E(N) = 1.5294$ and $\mathrm{Var}(N) = 4.8097$, respectively. Figure 7 shows graphs of the probability distribution and the cumulative probability distribution of the number of applications in the FS. Note that we chose $N = 25$ since at that point the cumulative has already been reached 1, as it should.

Table 5 Probabilities of the number of applications in FS, Example 10.3.3

P_0	P_1	P_2	P_3	P_4	P_5	P_6	P_7	P_8	P_9	P_{10}
0.4857	0.1388	0.1507	0.0748	0.0558	0.0330	0.0222	0.0139	0.0109	0.0090	0.0058

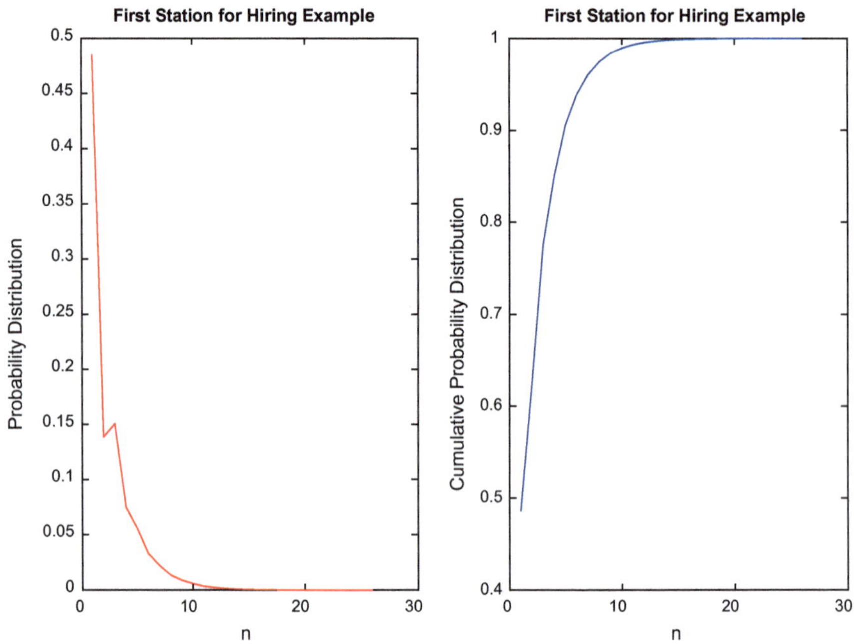

Fig. 7 Probabilities and cumulative probabilities of the number of applications in FS, Example 10.3.3

Example 10.3.4 In this example, we refer to Example 10.3.3 and assume the same distribution for the batch sizes, Bernoulli, and choose two different values for the batch sizes. We will then find the same measure as in the previous example. This would help the reader to watch for the steps necessary to obtain specific pdf of the number of applications in FS in case that the number of batch sizes increases. Thus, suppose that applications arrive in batches of size 2 or 3 according to Bernoulli distribution given by (10.3.18).

For this example, we denote the mean batch sizes, arrival rate, service rate, utilization factor, pgf for batch size distribution, and pgf for the distribution of the number of application, by $\bar{\alpha}$, λ, v ρ $G(z)$, and $A(z)$, respectively. Hence, $\bar{\alpha} = 2p - 3(1-p) = 3 - p$, ρ_2, and $A(z)$ are given as in (10.3.9), (10.3.10), and (10.3.11), respectively. Consequently, from (10.3.7) with the assumption that $b = B = 1$, similar to (10.3.21), we will have:

$$G(z) = \frac{1-\rho}{1 - \frac{\lambda}{v}z - \frac{\lambda}{v}z^2 - \frac{\lambda(1-p)}{v}z^3}. \tag{10.3.25}$$

Once again, from (10.3.3),

$$\rho = \frac{\lambda\bar{\alpha}}{v} = \frac{(3-p)\lambda}{v}. \tag{10.3.26}$$

Similar to (10.3.14), P_0 for this case is $P_0 = 1 - \rho$.

We should note that the value of P_0 could have been obtained from (10.3.25) by substituting $z = 0$ in it.

Using derivatives and evaluating them at $z = 0$, we can find the rest of the probabilities of the number of applications in the FS, two of which are as follows:

$$P_1 = \frac{\rho(1-\rho)}{3-\rho} \quad \text{and} \quad P_2 = \frac{\rho(1-\rho)(\rho+3-\rho)}{(3-p)^2}. \tag{10.3.27}$$

In general, we will have:

$$P_j = \frac{G^{(j)}(0)}{j!}, \quad j = 0, 1, 2, 3, \ldots. \tag{10.3.28}$$

The mean and variance of the number of applications in FS can, respectively, be found by taking derivatives, sequentially, of (10.3.25) and evaluating at $z = 1$. Hence, we have:

$$E(N) = \frac{3\rho(2-p)}{(1-p)(3-p)} \tag{10.3.29}$$

and

$$\mathrm{Var}(N) = \frac{9\rho^2(2-p)^2 + 2\rho(1-\rho)(3-\rho)(4-3\rho)}{(1-\rho)^2(3-\rho)^2}. \tag{10.3.30}$$

If $p = 2/3$, and we choose $\lambda = 3$ and $v = 8$, then for $N = 0, 1\ldots, 135$, we obtain $\rho_2 = 0.875$, $P_0 = 1 - \rho = 1 - 0.875 = 0.125$, $P_1 = 0.0515$, $P_2 = 0.065$, $E(N) = 4.5$, and $\mathrm{Var}(N) = 156$. Thus, the probabilities of having 0 up to 10 applications in the FS are listed in Table 6. Additionally, the mean and variance of the number of applications are $E(N) = 1.5294$ and $\mathrm{Var}(N) = 4.8097$, respectively. Figure 8 shows graphs of the probability distribution and the cumulative probability distribution of the number of applications in the FS. Note that we chose $N = 135$ since at that point the cumulative has already been reached 1, as it should.

Table 6 Probabilities of the number of applications in FS, Example 10.3.4

P_0	P_1	P_2	P_3	P_4	P_5	P_6	P_7	P_8	P_9	P_{10}
0.1250	0.0469	0.0645	0.0574	0.0515	0.0489	0.0448	0.0416	0.0385	0.0356	0.0330

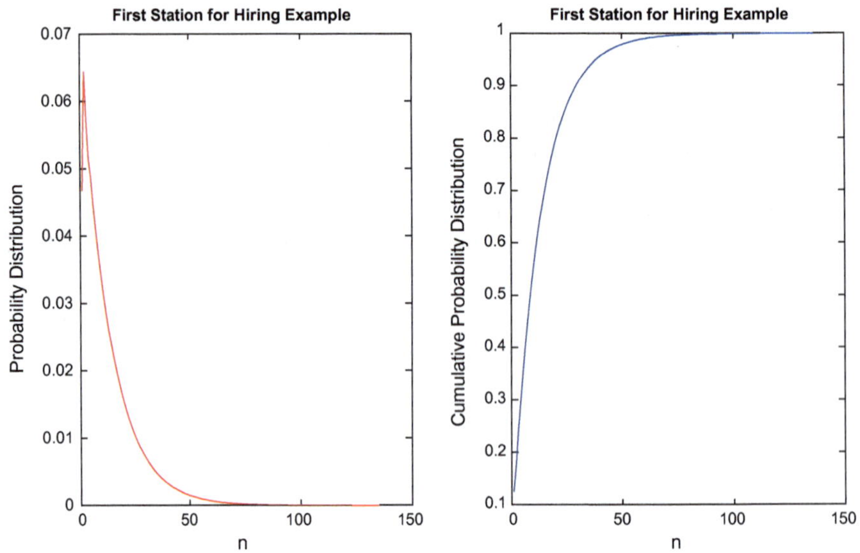

Fig. 8 Probability and cumulative probability distributions of the number of applications in FS, Example 10.3.4

10.3.3 Number of Batches in the FS at a Departure of a Batch

We now return to the general case. Let us denote the number of batches of application in the FS at the departure of the nth batch from the FS by Y_n. Let us also suppose that initially there are Y_0 batches of applications in the Inbox waiting for their processes to begin. Thus, the departure point process $\{Y_n : n \geq 1\}$ is a Markov chain, which is embedded over the non-Markovian process at regeneration points. Hence, we let ξ_n denote the random variable representing the number of batches arrived during the processing time of the nth batch. We finally denote the distribution of ξ_n by

$$d_m \equiv P(\xi_n = m), \quad m = 0, 1, \ldots. \tag{10.3.31}$$

We now consider possible cases at a departure point. Let q denote the number of applications left behind at the departure of the nth batch and obtain the following as the summary:

$$Y_{n+1} = \begin{cases} \xi_{n+1}, & \text{if } Y_n = 0, \\ Y_n - q + \xi_{n+1}, & \text{if } Y_n > 0,\ b \leq q < B, \\ Y_n - B + \xi_{n+1}, & \text{if } Y_n > 0,\ q \geq B. \end{cases} \tag{10.3.32}$$

Based on our assumption that arrival of applications follows a Poisson process with parameter $\lambda\alpha_s, 0 \le \alpha_s \le 1, s = 1, 2, \ldots$, the conditional distribution of ξ_n given the processing time $\Theta = \theta$ will be

$$P(\xi_n = m | \Theta = \theta) = \frac{(\lambda\alpha_x\theta)^m e^{-\lambda\alpha_s\theta}}{m!}, \quad 0 \le \alpha_s \le 1, \; m = 0, 1, 2, \ldots, \; s = 1, 2, \ldots. \tag{10.3.33}$$

Therefore,

$$\begin{aligned} d_m = P(\xi_n = m) &= \int_0^\infty \frac{(\lambda\alpha_s\theta)^m}{m!} e^{-\lambda\alpha_s\theta} \nu e^{-\nu\theta} d\theta \\ &= \frac{\nu(\lambda\alpha_s)^m}{m!} \int_0^\infty \theta^m e^{-(\lambda\alpha_s + \nu)\theta} d\theta, \quad 0 \le \alpha_x \le 1, \; m = 0, 1, 2, \ldots, \; s = 1, 2, \ldots. \end{aligned} \tag{10.3.34}$$

Denoting the transition probabilities of the Markov chain $\{Y_n : n \ge 1\}$ by P_{ij},

$$p_{km} = P(Y_{n+1} = m | Y_n = k), \tag{10.3.35}$$

from (10.3.34) and (10.3.35) we will have

$$p_{km} = \begin{cases} d_{m-k+B}, & \text{if } k > B, \; m \ge k - B, \\ d_m, & \text{if } 0 \le k \le B, \; m \ge 0, \\ 0, & \text{if } k \ge B, \; m < k - B. \end{cases} \tag{10.3.36}$$

Writing the transition matrix of (10.34), say **P**, which would be an *upper Hessenberg* matrix and since the first B + 1 rows of this matrix are identical, the Markov chain represented by the matrix is irreducible and periodic. We denote by $\prod_n$ ***the steady-state probability of n batches of applications in the FS at a departure time.*** We also assume that the traffic intensity $\rho = \frac{\lambda\bar{\alpha}}{B\nu} < 1$, according to Jain et al. [15]. Then the matrix **P** is ergotic. Hence, $\{\prod_n : n \ge 0\}$ exists. Thus, we can find $\prod_n$ from the system of balance equations,

$$\prod_m = d_m \sum_{i=0}^{B-1} \prod_i + \sum_{i=B}^{B+m} d_{m-i+B} \prod_i, \quad m = 0, 1, \ldots \tag{10.3.37}$$

with $\sum_{m=0}^{\infty} \prod_m = 1$.

Denoting the generating function of $\prod_m$ by $\prod(z)$ and using the generating function method, $\prod(z)$ will be found. Thus, let the generating function of d_m, denoted by $\Delta(z)$, be defined as:

$$\Delta(z) = \sum_{m=0}^{\infty} d_m z^m, \tag{10.3.38}$$

where $\Delta(z)$ is a function of both z and x and since $\sum_{m=0}^{\infty} d_m = 1$, $\Delta(1) = 1$, where $\mathbf{1}$ is a matrix with all elements as 1. Then, for $b = B = 1$, we will have

$$\prod(z) = \frac{(z-1)\Delta(z)}{z - \Delta(z)}(1-\rho), \tag{10.3.39}$$

which is the Polaczek-Khinchin formula; and for $B \geq 2$ we will have:

$$\prod(z) = \frac{\Delta(z)\sum_{i=0}^{B-1} \prod_i (z^B - z^i)}{z^B - \Delta(z)}, \quad |z| \leq 1, \tag{10.3.40}$$

or

$$\prod(z) = \frac{(B-\rho)(z-1)\Delta(z)\prod_{i=1}^{B-1}\left(\frac{z-r_i}{1-r_i}\right)}{z^B - \Delta(z)}, \quad |z| \leq 1, \ \rho = \frac{\lambda \bar{a}}{B\nu} < 1, \tag{10.3.41}$$

where $1, r_1, r_2, \ldots, r_{B-1}$ are roots of the denominator of $z^B - \Delta(z)$ inside and on the unit circle. See Haghighi and Mishev [3, 10]. Note that p_0 may be found as $\prod(0)$.

To find the roots in the unit circle of the denominator of (10.3.41), in addition to other methods available in the literature, we may do as follows in MATLAB computer software, say version (2017a):

1. Combine z^B with the first column of polynomial $\Delta(z)$,
2. Put the coefficients of $z^B - \Delta(z)$ in a vector, say $C_\Delta = (C_1\, C_2\, C_3\, C_4\, C_5\, C_6\, C_7)$,
3. Write $roots_\Delta(C_\Delta)$, and
4. Choose real roots that are less than 1.

To solve for $\prod(z)$, Haghighi and Mishev [3, 10] have offered an algorithm and an example.

Example 10.3.5 We assume that the batch size random variable X has a geometric distribution as follows:

$$\alpha_s = P(X = x) = p(1-p)^{s-1}, \quad 0 < p < 1, \ s = 1, 2, \ldots. \tag{10.3.42}$$

with parameter p. Of course the mean and generating function for this distribution are $1/p$ and

$$A(z) = \frac{pz}{1-(1-p)z}, \tag{10.3.43}$$

respectively. We use the following data: $\lambda = 3, \nu = 10, p = 0.4$, and $B = 1$. Hence, $\rho = 0.75$ and from (10.3.41), $p_0 = 0.25$. From the given data, we can see that for $s = 1, 2, \ldots, 20$, values of α_x are shown in Table 7.

If we chose $B = 1$, $M = 6$, and infinity as 6, then elements of the first column of matrix d_m, for $m = 0, 1, 2, 3, 4, 5, 6$, from (10.3.34), taking infinity as 20 to make the sum of probability for each x to be 1, which gives the probability of the number of batches arrived during a service for batches of different sizes, and then the first element of the vector $\Delta(z)$ from (10.3.38) will, respectively, be as follows:

$$d_m = (0.8929, 0.0957, 0.0102, 0.0011, 0.0001, 0.0000, 0.0000),$$
$$\Delta(z) = 0.0000013507z^6 + 0.000012607z^5 + 0.00011766z^4 + 0.0011z^3 + 0.0102z^2 + 0.0957z + 0.8929$$

10.4 A Queueing System with Delayed Feedback

In this subsection, we consider a different model that has application in our daily lives. Delays occur in many different instances such as traffic, signal outrage, and computer breakage. In many cases, time is needed for traffic to fellow normally, for signal to be connected and computer is repaired. The time we mentioned is delay time. However, the model we will consider is the one where delay occurs on returning to the server after the departed job is observed to need further service.

The study of queueing systems with immediate Bernoulli feedback was introduced by Takács [16]. Montazer-Haghighi [7] (one of Takács's doctoral students at Case Western Reserve University in Cleveland, Ohio, USA) extended the systems as a multichannel service, as well as a brief discussion of delayed feedback without offering a solution. Later, Disney and Kiessler [17] considered queues with delayed feedback as a traffic process. Kleinrock and Gail [18] discussed an *M*/*M*/1 with random delayed feedback. Haghighi et al. [19] considered a similar queue, but with an additional feature, a splitting device set that a serviced task might choose to go through after the service station. A detailed discussion of this appeared in Haghighi and Mishev [20]. Haghighi et al. [21] discussed a delayed service model that was detailed in Haghighi and Mishev [3] as an *M*/*G*/1 queueing system. The processing time for this system consisted of two

Table 7 Probabilities of the number of applications in a batch, Example 10.3.5

α_1	α_2	α_3	α_4	α_5	α_6	α_7	α_8	α_9	α_{10}
0.4000	0.2400	0.1440	0.0864	0.0518	0.0311	0.0187	0.0112	0.0067	0.0040
α_{11}	α_{12}	α_{13}	α_{14}	α_{15}	α_{16}	α_{17}	α_{18}	α_{19}	α_{20}
0.0024	0.0015	0.0009	0.0005	0.0003	0.0002	0.0001	0.0001	0.0000	0.000

independent parts, the delay time and the service time.Hungarian Mathematician, Lajos Takács (August 21, 1924–December 4, 2015 From: https://www.pvamu.edu/sites/mathematics/journal/aam/2015/vol-10-issue-2/01_AAM_Vol_10_Issue_2_Preface_Ready_AMH.pdf

10.4.1 The Model

Our model illustrated in Fig. 9 is a special case of Haghighi and Mishev [20] that also briefly appeared in Haghighi and Mishev [22].

Jobs singly arrive from an infinite source outside the system according to a Poisson distribution with parameter λ. There is a single server with an infinite size buffer in front of it. When a job arrives and finds the server busy, it will join the waiting line in the buffer. The service distribution is exponential with parameter μ. Service is on basis of FCFS. After a job leaves the server, it will either leave the system forever with probability $q, 0 \leq q \leq 1$; or will return to the buffer, at the end of the waiting line, for further service with probability $p, 0 \leq p \leq 1$; $q + p = 1$. However, the feedback to the service station is not immediate. There is another

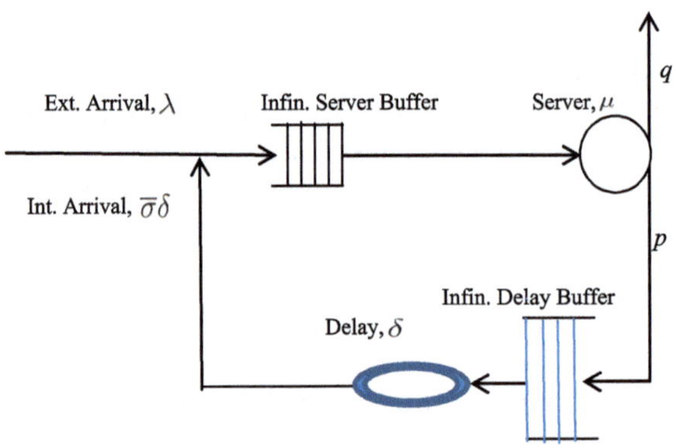

Fig. 9 Single-processor with infinite server-buffer and delayed batch Feedback

infinite buffer, referred to as the *delay-station*, between the server and the first buffer such that a returnee job must go through it to get back to the service station again. A *mover* moves the jobs entered into the delay-station.

Thus, in reality, there are two kinds of arrivals: external arrivals from outside and internal arrivals that are returnees. There is a processing procedure in the delay-station. Jobs in the delay-station group with varying random sizes between two natural numbers minimum b and maximum B, $1 \leq b \leq B$. Thus, the group (batch) sizes may be represented by random variables, denoted by X, with probability distribution function and values between b and B, inclusive, as:

$$P\{X = x\} = \sigma_x, \quad b \leq x \leq B, \tag{10.4.1}$$

with mean random size of a group denoted by σ. We assume that both the mean and the variance of X are positive and finite. We also assume that the delay times have exponential distribution with parameter δ. Based on the property of the exponential distribution mentioned earlier, internal arriving jobs arrive from the delay-station by groups that follow a Poisson distribution with parameter δ batches per unit time. Hence, the average internal arriving job rate is $\bar{\sigma}\delta$ jobs per unit time and the rate into the service buffer will be $\lambda + \bar{\sigma}\delta$. The returnee groups open up at the service buffer and will be served one at a time according to their order of arrivals into the delayed station. We assumed that the return of a job is an event independent of any other event and, in particular, independent of the number of its returns.

We can think of the model described as a tandem queueing system, and our goal is to analyze the busy period of the server in its service station.

We will only discuss the busy period of the server for a special case of the queueing model discussed in Haghighi and Mishev [22].

10.4.2 Analysis

To begin with, we denote the expected attendance rates to the service-station, the traffic intensity (or the load) for the service station, and the same for the delay-station, by λ_1, ρ_1, λ_2, and ρ_2, respectively. Then, we will have:

$$\begin{cases} \lambda_1 = \lambda + \sigma\delta, \\ \lambda_2 = \lambda_1 p, \end{cases} \quad \text{and} \quad \rho_i \equiv \frac{\lambda_i}{\mu_i}, \; i = 1, 2, \tag{10.4.2}$$

where $\mu_1 = \mu$ and $\mu_2 = \bar{\sigma}\delta$, with each load restricted to be less than one. The intensity of the system or total load, denoted by ρ_{sys}, is

$$\rho_{sys} \equiv \frac{\lambda_1 + \lambda_2}{\mu + \bar{\sigma}\delta} = \frac{(\lambda + \bar{\sigma}\delta)(1 + p)}{\mu + \bar{\alpha}\delta}. \tag{10.4.3}$$

Of course we have to have $p_1 < 1, i = 1, 2$ and $p_{sys} < 1$ for the system to have finite distributions.

It should be reminded that both external and internal arrivals possess Poison distribution. The external is singly with rate λ and internal is by batch with rate $\bar{\sigma}\delta$. Thus, since the distribution of arrival is not immediately known, we can consider the service station as a *G*/*M*/1 (General arrival, Markovian service, and a single server) with service rate μ. It is well-known that *G*/*M*/1 and *M*/*G*/1 are dual-queue systems. Thus, we could solve either on. The service discipline is assumed to be first-come, first-served. In either model, the mean arrival rate is $1/(\lambda + \bar{\sigma}\delta)$ and the mean service time is $1/\mu$, regardless of the source of arrivals.

The literature is quite rich for busy period of a queue. For instance, (see Haghighi and Mishev [9], Haghighi et al. [21, 23]). Hence, what we will discuss here will be one of the many models discussing the busy period. However, we do that we now define a couple of terms we need soon. The *server's busy period* is the period starting from one state of the system to the immediate next state. We denote the *length of a busy period by B*. By an *idle period,* it is meant the period during which there is no job to be served. Combining these two periods, we will have a *busy cycle*.

We now assume that at time $t = 0$ the service stations at the arrival of the first job to the system at this time. Let us denote the pdf of a busy period, *Y*, by $f(t)$, defined $f(t) \equiv \phi_0'(t)$, where $\phi_0(t)$ is the probability of the service station being empty at time *t*, that is, $\phi_0(t)$ is the probability of $Y < t$. Hence,

$$f(t) = \frac{1}{t\sqrt{\rho_1}} e^{-(\lambda_1 + \mu)t} I_1\left(2t\sqrt{\lambda_1 \mu}\right), \quad t > 0, \tag{10.4.4}$$

where I_1 is the modified Bessel function of the first kind given by

$$I_r(x) = \sum_{m=0}^{\infty} \frac{1}{m!\Gamma(m+r+1)} \left(\frac{x}{2}\right)^{2m+r}, \tag{10.4.5}$$

Thus, distribution of the length of a busy period can be obtained by integrating (10.4.4) over the interval $(0, t]$. Thus, the mean and variance of a busy period, respectively, will be:

$$E\{Y\} = \frac{1}{\mu - \lambda_1} \quad \text{and} \quad \text{Var}\{Y\} = \frac{1 + \rho_1}{\mu^2 - (1 - \rho_1)^3}. \tag{10.4.6}$$

As it can be seen from (10.4.6), if $\mu \le \lambda_1$, then, the mean length of busy period can be infinite with some positive probability.

Let us now denote the length of an idle period by *I*. As we mentioned earlier, the interarrival times to the service station are exponentially distributed with parameter λ_1. Thus, the pdf of *I*, denoted by $\psi_I(t)$, is

$$\psi_I(t) = \lambda_1 e^{-\lambda_1 t}, \quad t > 0. \tag{10.4.7}$$

Example 10.4.1 Supposing that the distribution of the random variable representing the batch sizes, σ_x is a uniform distribution:

$$P\{X = x\} = \sigma_x = \begin{cases} \frac{x-b}{B-b}, & b \le x \le B, \\ 0, & \text{otherwise}. \end{cases} \tag{10.4.8}$$

Thus, we have

$$\bar{\sigma} = \frac{b+B}{2}. \tag{10.4.9}$$

Let us choose the following data: $\lambda = 20, \mu = 50, v = 15$. Thus, from (10.4.6), we will have $E(Y) = 1/30$ and $\text{Var}(Y) = 0.0000559$. For $b = 2$ and $B = 5$, the average size of a batch is 3.5.

References

1. Kolmogorov A (1931) Uber die analytischen methoden in der Wahrscheinli-chkeitsrechnung. Math Ann 104:415–458
2. Erlang AK (1917) Solution of some problems in the theory of probabilities of significance in automatic telephone exchanges. Elektrotkeknikeren 13
3. Haghighi AM, Mishev DP (2014) Queueing models in industry and business, 2nd edn. Nova Science Publishers Inc., New York
4. Haight FA (1957) Queueing with balking. Biometrika 44:360–369
5. Haight FA (1960) Queueing with balking II. Biometrika 47:285–296
6. Ancker CJ Jr, Gafarian AV (1963) Some queueing problems with balking and reneging I & II. Opns Res II(88–100):928–937
7. Haghighi AM (1976) Many-server queueing system with feedback. Doctoral dissertation, Case Western Reserve University, Cleveland, Ohio, USA
8. Haghighi AM, Medhi J, Mohanty SG (1986) On a multiserver Markovian queueing system with balking and reneging. Comput Oper Res 13(4):421–425
9. Haghighi AM, Mishev, DP (2006) A parallel priority queueing system with finite buffers. J Parallel Distrib Comput 66:379–392. http://www.sciencedirect.com/science/journal/07437315/66/3
10. Haghighi AM, Mishev DP (2013) Stochastic three-stage hiring model as a Tandem Queueing Process with bulk arrivals and Erlang phase-type selection, $M^X/M^{(k,K)}/1\text{-}M^Y/E_r/1\text{-}\infty$. Int J Math Oper Res (IJMOR) 5(5)
11. Haghighi AM, Mishev DP (2007) A Tandem Queueing system with task-splitting, feedback and blocking. Int J Oper Res (IJOR) 2(2):208–230
12. Neuts MF (1967) A general class of bulk queues with Poisson input. Ann Math Stat 38 (3):759–770. Original draft, Department of Statistics, Division of mathematical Sciences, Purdue University, Mimeograph Series No. 46, 1965
13. Madan KC (2003) An *M*/*G*/1 queue with time homogeneous breakdown and deterministic repair times. Soochow J Math 29
14. Medhi J (1991, 2003) Stochastic models in queueing theory, 2nd edn (2003). Academic Press
15. Jain JL, Mohanty SG, Böhm W (2007) A course on queueing models. Chapman & hall/CRC, Taylor & Francis Group

16. Takács L (1963) A single server queue with feedback. Bell Syst Tech J 42
17. Disney RL, Kiessler PC (1987) Traffic Processes in Queueing Networks: a Markov renewal approach. The John Hopkins University Press
18. Kleinrock L, Gail R (1996) Queueing systems, problems and solutions. Wiley, New York
19. Haghighi AM, Chukova S, Mishev DP (2011) Single-server Poisson queueing system with delayed-feedback: part 1. Int J Math Oper Res (IJMOR) 3(1)
20. Haghighi AM, Mishev DP (2013) Difference and differential equations with applications in Queueing theory. Wiley, Hoboken, New Jersey, United State of America and simultaneously in Canada
21. Haghighi AM, Chukova S, Mishev DP (2008) A two station Tandem Queueing system with delayed-service. Int J Oper Res 3(4)
22. Haghighi AM, Mishev DP (2016) Delayed and network queues. Wiley, New Jersey. http://www.wiley.com/WileyCDA/WileyTitle/productCd-1119022134,subjectCd-BA11.html
23. Haghighi AM, Lian J, Mishev DP (2011) Advanced mathematics for engineers with applications in stochastic processes. Nova Science Publishers Inc., New York

A Study on Optimal Preventive Maintenance Policies for Cumulative Damage Models

Naoto Kaio

1 Introduction

The preventive maintenance policy is one of the most important problems in the reliability theory and the maintenance one. The preventive maintenance models are classified into several categories, and many kinds of preventive maintenance policies have been discussed (e.g., [1–6, 10, 12–14]). Especially, there exists a preventive maintenance policy taking account of damage by shocks as one of them (e.g., [7, 9, 11]).

In this chapter, we consider the extended preventive maintenance policies for cumulative damage models with stochastic failure levels. That is, we discuss the optimal preventive maintenance policies for the system that fails when the cumulated amount of damage by shocks exceeds a stochastic failure level, assuming a continuous distribution and a discrete one, respectively. We apply the expected costs per unit time in the steady state as criteria of optimality and seek the optimal policies minimizing these expected costs. We show that there exists a unique optimal policy under certain conditions, respectively. Furthermore, we refer to the modified models where the shock does not always give the damage to the system.

N. Kaio (✉)
Department of Economic Informatics, Hiroshima Shudo University,
1-1-1 Ozukahigashi, Asaminami-ku, Hiroshima 731-3195, Japan
e-mail: kaio@shudo-u.ac.jp

M. Ram and J.P. Davim (eds.), *Modeling and Simulation in Industrial Engineering*,
Management and Industrial Engineering, DOI 10.1007/978-3-319-60432-9_8

2 Preventive Maintenance Policies for a Cumulative Damage Model with a Continuous Distribution

2.1 Model and Assumptions

1. Consider a one–unit system.
2. System failure is revealed at failure immediately and each failed unit is scrapped without repair.
3. An unlimited number of spare units are immediately available when they are needed.
4. The original unit begins operating at time 0 with the cumulated amount of damage 0.
5. The planning horizon is infinite.
6. The damage occurred by shocks is additive.
7. The unit fails only when the cumulated amount of damage by shocks exceeds the stochastic failure label r.v. (random variable) W ($W \geq 0$). The r.v. W obeys a cdf (cumulative distribution function) $D(w)$ ($w \geq 0$) with a pdf (probability density function) $d(w)$.
8. When the cumulated amount of damage exceeds the predetermined exchange level w_0 ($0 \leq w_0 < \infty$) by any shock, the unit is exchanged if it has not failed, by the spare one immediately (i.e., the preventive maintenance). On the other hand, the unit is replaced if it has failed by any shock in a similar fashion (i.e., the corrective one).
9. The exchange and replacement actions are executed instantaneously. The new exchanged and replaced units take over the operation immediately.
10. The similar cycles are repeated from time to time. That is, an interval from the start of the original unit (the exchange or the replacement) to the following exchange or replacement is defined as one cycle, and the cycle repeats itself again and again.
11. The time interval between $j-1$st shock and jth one is r.v. T_j ($j = 1, 2, 3, \ldots$; $T_j \geq 0$ and r.v. T_1 is the time interval between the time 0 and the first shock), and the amount of damage by jth shock is r.v. X_j ($X_j \geq 0$), where r.v. X_i is independent of r.v. T_j ($i \neq j$).
12. There exist n types of shock modes and the shock mode i occurs with probability a_i ($i = 1, 2, \ldots, n$; $\sum_{i=1}^{n} a_i = 1$; $a_i \geq 0$). Under the condition that the mode of jth shock is the ith type, r.v. T_j obeys a cdf $F_i(t)$ ($t \geq 0$), and r.v. X_j obeys a cdf $G_i(x)$ ($x \geq 0$), that is, to say r.v. T_j obeys the cdf $F(t) = \sum_{i=1}^{n} a_i F_i(t)$, and r.v. X_j obeys the cdf $G(x) = \sum_{i=1}^{n} a_i G_i(x)$. Furthermore, we put $\int_0^{\infty} t dF_i(t) = 1/\lambda_i$ and $\int_0^{\infty} t dF(t) = 1/\lambda$ and this implies the relation $1/\lambda = \sum_{i=1}^{n} a_i/\lambda_i$. The amount of damage by jth shock, r.v. X_j has a renewal function $M(x)$ and a renewal density $m(x)$.

13. The costs considered are a cost c_0 suffered for each exchange before a failure (each preventive maintenance) and a cost c_i suffered for each replacement after a failure (each corrective one) due to shock mode i, where $c_i > c_0$ since the corrective maintenance is more expensive than the preventive one.

Under these model and assumptions, we derive the expected cost per unit time in the steady state and discuss the optimal preventive maintenance policies minimizing that expected cost.

2.2 Analysis and Theorems

The expected cost per one cycle $A_c(w_0)$ is given by

$$
\begin{aligned}
A_c(w_0) = c_0 \int_{w_0}^{\infty} [G(w) - \int_0^{w_0} \bar{G}(w-u)\mathrm{d}M(u)]dD(w) \\
+ \sum_{i=1}^{n} a_i c_i \left[\int_0^{w_0} \left\{ \bar{G}_i(w) + \int_0^{w} \bar{G}_i(w-u)\mathrm{d}M(u) \right\} \mathrm{d}D(w) \right. \\
\left. + \int_{w_0}^{\infty} \left\{ \bar{G}_i(w) + \int_0^{w_0} \bar{G}_i(w-u)\mathrm{d}M(u) \right\} \mathrm{d}D(w) \right],
\end{aligned}
\tag{1}
$$

where $\bar{\psi}(\cdot) = 1 - \psi(\cdot)$, in general.

The mean time of one cycle $B_c(w_0)$ is

$$
B_c(w_0) = (1/\lambda) \left[\int_0^{w_0} \{1 + M(w)\}\mathrm{d}D(w) + \{1 + M(w_0)\}\bar{D}(w_0) \right]. \tag{2}
$$

We obtain the following when $w_0 = 0$ and $w_0 \to \infty$.

$$
A_c(0) = c_0 \int_0^{\infty} G(w)\mathrm{d}D(w) + \sum_{i=1}^{n} a_i c_i \int_0^{\infty} \bar{G}_i(w)\mathrm{d}D(w) , \tag{3}
$$

$$
B_c(0) = 1/\lambda , \tag{4}
$$

$$
A_c(\infty) = \sum_{i=1}^{n} a_i c_i \int_0^{\infty} \left[\bar{G}_i(w) + \int_0^{w} \bar{G}_i(w-u)\mathrm{d}M(u) \right] \mathrm{d}D(w) , \tag{5}
$$

and

$$B_c(\infty) = (1/\lambda)\int_0^\infty [1+M(w)]\mathrm{d}D(w)\,. \tag{6}$$

Thus, the expected cost per unit time in the steady state is given by

$$C_c(w_0) = \frac{A_c(w_0)}{B_c(w_0)} \tag{7}$$

(see [15], p. 52).

Define the numerator divided by $m(w_0)\bar{D}(w_0)$ of the derivative of the right-hand side in Eq. (7) as

$$\begin{aligned} q_c(w_0) = [1/\bar{D}(w_0)]&\Bigg[-c_0\int_{w_0}^\infty \bar{G}(w-w_0)\mathrm{d}D(w) \\ &+ \sum_{i=1}^n a_i c_i \int_{w_0}^\infty \bar{G}_i(w-w_0)\mathrm{d}D(w)\Bigg]B_c(w_0) \\ &- A_c(w_0)(1/\lambda), \end{aligned} \tag{8}$$

where $q_c(0) = -c_0/\lambda < 0$.

We obtain the following theorems with respect to the optimal exchange level w_0^* minimizing the expected cost per unit time in the steady state $C_c(w_0)$ in Eq. (7).

Theorem 1 *There exists at least one positive optimal exchange level* w_0^* $(0 < w_0^* \leq \infty)$. *If* $q_c(\infty) > 0$, *then there exists at least one positive and finite optimal exchange level* w_0^* $(0 < w_0^* < \infty)$.

Proof These results hold clearly since $q_c(0) = -c_0/\lambda < 0$. □

Theorem 2

1. *Suppose that* $q_c(w_0)$ *is strictly increasing.*

(i) *If* $q_c(\infty) > 0$, *then there exists a finite and unique optimal exchange level* w_0^* $(0 < w_0^* < \infty)$ *satisfying* $q_c(w_0) = 0$ *and the corresponding expected cost is*

$$\begin{aligned} C_c(w_0^*) = [\lambda/\bar{D}(w_0^*)]&\Bigg[-c_0\int_{w_0^*}^\infty \bar{G}(w-w_0^*)\mathrm{d}D(w) \\ &+ \sum_{i=1}^n a_i c_i \int_{w_0^*}^\infty \bar{G}_i(w-w_0^*)\mathrm{d}D(w)\Bigg]. \end{aligned} \tag{9}$$

(ii) *If* $q_c(\infty) \leq 0$, *then the optimal exchange level is* $w_0^* \to \infty$, *i.e., the unit continues operation until its failure and then it is replaced by the new one, and the corresponding expected cost is* $C_c(\infty) = A_c(\infty)/B_c(\infty)$.

2. *When* $q_c(w_0)$ *is decreasing, we have* $w_0^* \to \infty$.

Proof These results hold clearly from the monotone properties of $q_c(w_0)$ and Theorem 1. □

2.3 Remarks

In the present model, if we put $c_N = c_i$ $(i = 1, 2, \ldots, n)$, $D(w) = u(w - W_0)$ (unit function), and $w_0 < W_0$, or if we put $c_N = \sum_{i=1}^{n} a_i c_i$, $G(x) = G_i(x)$, $D(w) = u(w - W_0)$, and $w_0 < W_0$, we have

$$A_c(w_0) = c_0 + (c_N - c_0)\left[\bar{G}(W_0) + \int_0^{w_0} \bar{G}(W_0 - u)\mathrm{d}M(u)\right], \tag{10}$$

and

$$B_c(w_0) = [1 + M(w_0)]/\lambda. \tag{11}$$

This is equivalent to the result discussed by Nakagawa [9].

In the present model, we assume that the shock gives the damage to the system with probability 1. Next, we refer to the preventive maintenance policy for the modified continuous type cumulative damage model, where the shock does not always give the damage to the system (see [11]). That is, we consider the situation that the shock generates the damage to the system with probability p $(0 < p \leq 1)$, i.e., it does not with probability $1 - p$. Also for this situation, we can apply our results by using λp instead of λ.

3 Preventive Maintenance Policies for a Cumulative Damage Model with a Discrete Distribution

3.1 Model and Assumptions

We apply items 1–6, 9, 10, and 13 not only in Sect. 2.1 but also in this section, and we rewrite items 7, 8, 11, and 12 for the discrete distribution as follows.

7. The unit fails only when the cumulated amount of damage by shocks exceeds the stochastic failure label r.v. V ($V = 0, 1, 2, \ldots$). The r.v. V obeys a cdf $K(v)$ ($v = 0, 1, 2, \ldots$) with a pmf (probability mass function) $k(v)$ ($k(0) = 0$).
8. When the cumulated amount of damage exceeds the predetermined exchange level v_0 ($v_0 = 0, 1, 2, \ldots$) by any shock, the unit is exchanged if it has not failed, by the spare one immediately (the preventive maintenance). On the other hand, the unit is replaced if it has failed by any shock in a similar fashion (the corrective one).
11. The time interval between $k-1$st shock and kth one is r.v. B_k ($k = 1, 2, 3, \ldots$; $B_k = 0, 1, 2, \ldots$ and r.v. B_1 is the time interval between the time 0 and the first shock), and the amount of damage by kth shock is r.v. D_k ($D_k = 0, 1, 2, \ldots$), where r.v. D_i is independent of r.v. B_j ($i \neq j$).
12. There exist n types of shock modes and the shock mode i occurs with probability a_i ($i = 1, 2, \ldots, n; \sum_{i=1}^{n} a_i = 1; a_i \geq 0$). Under the condition that the mode of kth shock is the ith type, r.v. B_k obeys a cdf $F_i(b)$ ($b = 0, 1, 2, \ldots$) with a pmf $f_i(b)$ ($f_i(0) = 0$), and r.v. D_k obeys a cdf $G_i(d)$ ($d = 0, 1, 2, \ldots$), that is, r. v. B_k obeys the cdf $F(b) = \sum_{i=1}^{n} a_i F_i(b)$ with the pmf $f(b) = \sum_{i=1}^{n} a_i f_i(b)$, and r.v. D_k obeys the cdf $G(d) = \sum_{i=1}^{n} a_i G_i(d)$. Furthermore, we put $\sum_{b=0}^{\infty} b f_i(b) = 1/\lambda_i$ and $\sum_{b=0}^{\infty} b f(b) = 1/\lambda$ and this implies the relation $1/\lambda = \sum_{i=1}^{n} a_i/\lambda_i$. The amount of damage by kth shock, r.v. D_k has a renewal function $M(d)$ and a renewal probability mass function $m(d)$ (see [8]).

Under these model and assumptions, we derive the expected cost per unit time in the steady state and discuss the optimal preventive maintenance policies minimizing that expected cost.

3.2 Analysis and Theorems

The expected cost per one cycle $A_d(v_0)$ is given by

$$\begin{aligned} A_d(v_0) = c_0 \sum_{v=v_0+1}^{\infty} [G(v) - \sum_{l=0}^{v_0} \bar{G}(v-l)m(l)]k(v) \\ + \sum_{i=1}^{n} a_i c_i [\sum_{v=0}^{v_0} \{\bar{G}_i(v) + \sum_{l=0}^{v} \bar{G}_i(v-l)m(l)\}k(v) \\ + \sum_{v=v_0+1}^{\infty} \{\bar{G}_i(v) + \sum_{l=0}^{v_0} \bar{G}_i(v-l)m(l)\}k(v)]. \end{aligned} \tag{12}$$

The mean time of one cycle $B_d(v_0)$ is

$$B_d(v_0) = (1/\lambda)\left[\sum_{v=0}^{v_0} \{1 + M(v)\}k(v) + \{1 + M(v_0)\}\bar{K}(v_0)\right]. \tag{13}$$

We obtain the following when $v_0 = 0$ and $v_0 \to \infty$.

$$A_d(0) = c_0 \sum_{v=1}^{\infty} G(v)k(v) + \sum_{i=1}^{n} a_i c_i \sum_{v=1}^{\infty} \bar{G}_i(v)k(v) \, , \tag{14}$$

$$B_d(0) = 1/\lambda \, , \tag{15}$$

$$A_d(\infty) = \sum_{i=1}^{n} a_i c_i \sum_{v=0}^{\infty} [\bar{G}_i(v) + \sum_{l=0}^{v} \bar{G}_i(v-l)m(l)]k(v) \, , \tag{16}$$

and

$$B_d(\infty) = (1/\lambda) \sum_{v=0}^{\infty} [1 + M(v)]k(v) \, . \tag{17}$$

Thus, the expected cost per unit time in the steady state is given by

$$C_d(v_0) = \frac{A_d(v_0)}{B_d(v_0)} \tag{18}$$

(see [15], p. 52).

Define the numerator divided by $m(v_0+1)\bar{K}(v_0)$ of the difference of $C_d(v_0)$ in Eq. (18) as

$$\begin{aligned} q_d(v_0) = [1/\bar{K}(v_0)] \Bigg[& -c_0 \sum_{v=v_0+1}^{\infty} \bar{G}(v - v_0 - 1)k(v) \\ & + \sum_{i=1}^{n} a_i c_i \sum_{v=v_0+1}^{\infty} \bar{G}_i(v - v_0 - 1)k(v) \Bigg] B_d(v_0) \\ & - A_d(v_0)(1/\lambda). \end{aligned} \tag{19}$$

We obtain the following theorems with respect to the optimal exchange level v_0^* minimizing the expected cost per unit time in the steady state $C_d(v_0)$ in Eq. (18).

Theorem 3

1. *If* $q_d(\infty) > 0$, *then there exists at least one finite optimal exchange level* v_0^* $(0 \le v_0^* < \infty)$.
2. *If* $q_d(0) < 0$, *then there exists at least one positive optimal exchange level* v_0^* $(0 < v_0^* \le \infty)$.

Proof These results hold clearly. □

Theorem 4

1. *Suppose that* $q_d(v_0)$ *is strictly increasing.*

   (i) *If* $q_d(0) < 0$ *and* $q_d(\infty) > 0$, *then there exists a finite and unique optimal exchange level* v_0^* ($0 < v_0^* < \infty$) *satisfying* $q_d(v_0 - 1) < 0$ *and* $q_d(v_0) \geq 0$. *We have the following relationship with respect to the optimal expected cost.*

$$\begin{aligned}
&[\lambda/\bar{K}(v_0^* - 1)]\left[-c_0 \sum_{v=v_0^*}^{\infty} \bar{G}(v - v_0^*)k(v)\right.\\
&\left. + \sum_{i=1}^{n} a_i c_i \sum_{v=v_0^*}^{\infty} \bar{G}_i(v - v_0^*)k(v)\right]\\
&< C_d(v_0^*)\\
&\leq [\lambda/\bar{K}(v_0^*)]\left[-c_0 \sum_{v=v_0^*+1}^{\infty} \bar{G}(v - v_0^* - 1)k(v)\right.\\
&\left. + \sum_{i=1}^{n} a_i c_i \sum_{v=v_0^*+1}^{\infty} \bar{G}_i(v - v_0^* - 1)k(v)\right].
\end{aligned} \tag{20}$$

   (ii) *If* $q_d(\infty) \leq 0$, *then the optimal exchange level is* $v_0^* \to \infty$, *and the corresponding expected cost is* $C_d(\infty) = A_d(\infty)/B_d(\infty)$.

   (iii) *If* $q_d(0) \geq 0$, *then the optimal exchange level is* $v_0^* = 0$, *i.e., the unit is exchanged or replaced (failure) by the new one at the first shock. The corresponding expected cost is* $C_d(0) = A_d(0)/B_d(0)$.

2. *When* $q_d(v_0)$ *is decreasing, we have* $v_0^* \to \infty$ *or* $v_0^* = 0$.

Proof These results hold clearly from the monotone properties of $q_d(v_0)$ and Theorem 3. □

3.3 Remarks

We put

$$u(d) = \sum_{j=0}^{d} \delta(j) = 1, d = 0, 1, 2, \ldots, \tag{21}$$

where

$$\delta(d) = \begin{cases} 1, & d = 0, \\ 0, & d = 1, 2, 3, \ldots. \end{cases} \tag{22}$$

In the present model, if we put $c_N = c_i$ $(i = 1, 2, \ldots, n)$, $K(v) = u(v - V_0)$, and $v_0 < V_0$, or if we put $c_N = \sum_{i=1}^{n} a_i c_i$, $G(d) = G_i(d)$, $K(v) = u(v - V_0)$, and $v_0 < V_0$,

$$A_d(v_0) = c_0 + (c_N - c_0)[\bar{G}(V_0) + \sum_{l=0}^{v_0} \bar{G}(V_0 - l)m(l)], \tag{23}$$

and

$$B_d(v_0) = [1 + M(v_0)]/\lambda. \tag{24}$$

This is equivalent to the result discussed by Kaio and Osaki [7].

In the similar fashion of Sect. 2.3, when the shock generates the damage to the system with probability p $(0 < p \leq 1)$, i.e., it does not with probability $1 - p$, we can apply our results by using λp instead of λ.

4 Concluding Remarks

In this chapter, we have discussed the preventive maintenance policies for the extended cumulative damage model, in which the system fails when the cumulated amount of damage by shocks exceeds a stochastic failure level, assuming several shock modes, and a continuous distribution and a discrete one, respectively. We have applied the expected costs per unit time in the steady state as criteria of optimality and sought the optimal policies minimizing these expected costs. We also have shown the relationships between the results of this chapter and ones obtained in the earlier contributions.

References

1. Barlow RE (1998) Engineering reliability, ASA–SIAM, Philadelphia
2. Barlow RE, Proschan F (1965) Mathematical theory of reliability. Wiley, New York
3. Barlow RE, Proschan F (1981) Statistical theory of reliability and life testing: probability models. TO BEGIN WITH, Silver Spring, MD
4. Ben-Daya M, Duffuaa SO, Raouf A (eds) (2000) Maintenance, modeling and optimization. Kluwer Academic Publisher, Norwell, MA
5. Dohi T, Nakagawa T (eds) (2013) Stochastic reliability and maintenance modeling. Springer, London

6. Hayakawa Y, Irony T, Xie M (eds) (2001) System and Bayesian reliability. World Scientific, Singapore
7. Kaio N, Osaki S (1985) Optimum replacement policies for a discrete cumulative damage model (in Japanese). Trans IEICE Japan, J68–A, 981–982
8. Kaio N, Osaki S (1988) Review of discrete and continuous distributions in replacement models. Int J Syst Sci 19:171–177
9. Nakagawa T (1976) On a replacement problem of a cumulative damage model. Opl Res Q, 27:895–900
10. Nakagawa T (2005) Maintenance theory of reliability. Springer, London
11. Nakagawa T, Osaki S (1974) Some aspects of damage models. Microelectron Reliab 13:253–257
12. Osaki S (1985) Stochastic system reliability modeling. World Scientific, Singapore
13. Osaki S (1992) Applied stochastic system modeling. Springer, Berlin
14. Pham H (ed) (2003) Handbook of reliability engineering. Springer, London
15. Ross SM (1970) Applied probability models with optimization applications. Holden-Day, San Francisco

Elevator System Analysis in Deliberation of Dependability, Cost Under Coverage, and Copula Approaches

Nupur Goyal, Akshat Tandon, Anshumaan Saxena, Mangey Ram and J. Paulo Davim

1 Introduction

An elevator is a platform, either open or enclosed, used for lifting people or freight to upper floors within a building. Elevators are a standard part of any tall commercial or residential building. In recent years, the introduction of the Federal Americans with Disabilities Act has required that two- and three-story building be retrofitted with elevators. Elevators themselves are simple devices, and a basic lifting system which has not changed much in over 50 years. However, the control system has changed substantially to improve safety and speed of operation [1].

Elevators are designed for a specific building, taking into account such factors as the height of the building, the number of people traveling to each floor, and the expected periods of high usage. The most popular elevator designed is the roped elevators, where a car is raised and attached by steel cables. Here, the machine muscle lives up at the top of the elevator shaft, its ropes attached to the cars, and loops around the pulley with groups to grip the ropes that connected to electric

N. Goyal · M. Ram (✉)
Department of Mathematics, Graphic Era University, Dehradun, Uttarakhand, India
e-mail: drmrswami@yahoo.com

N. Goyal
e-mail: nupurgoyalgeu@gmail.com

A. Tandon · A. Saxena
Department of Mechanical Engineering, Graphic Era University, Dehradun, Uttarakhand, India
e-mail: tendon.akshat2612@gmail.com

A. Saxena
e-mail: saxena.ams@gmail.com

J.P. Davim
Department of Mechanical Engineering, University of Aveiro, Aveiro, Portugal
e-mail: pdavim@ua.pt

M. Ram and J.P. Davim (eds.), *Modeling and Simulation in Industrial Engineering*, Management and Industrial Engineering, DOI 10.1007/978-3-319-60432-9_9

motors. Turning the motor one way, the elevator car goes up when the motors on the other way, it goes down. A counterweight lives on the other end of the rope to offset the weight of the car. It is usually about the half of the weight of the freely loaded passenger of elevators. The counterweight is the most important part of the elevator; it provides force to the car for the motion. So, both are perfectly balanced on an average ride. This system saves energy as well as wear and tear of moving parts [2].

Once the car is moved by the motor, its only job is to control one of the falling objects. Both the car and counterweight are attached to guide rails inside the shaft. They keep everything from swing back and forth, and also give a backup setup of brakes, something to grab onto. If anything goes wrong with the motor, hydraulic fluid is cutoff and it automatically releases the brakes that cease the rope for quick stop [1]. Technically, one of the steel ropes is enough to hold the car and the counterweight, and rests of these are for backup in case one snaps. But, if the whole set is cut, the machine has a built-in brake set to stop the elevator.

The governor that engages with the motor with its own pulley and separate cables attached to the car. There are two spring-loaded naval hooks called fly weights inside the governor. Freely falling cabin results in faster motion of governor that pushes the hook out due to centrifugal force. The governor ropes jerk on an arm on top of the car, and this locks the brake [3].

To find out the location and instant, where and when the car stops, a call button is used in the elevator's car which sends signals to its operating system. It keeps track of every car in every shaft through sensors on the shaft wall. The weight of the elevator car is measure by mean of logical tracking number system. Then, it will skip your floor and come back when it starts. The idea of rope elevator is simple enough, one side goes up and the other goes down.

Recently, an elevator is one of the most desired requirements of every organization in every field of industry, but rope elevator has wide application in every organization. The main function of an elevator is to transport the necessary components and objects to the required destination. In the present era of system technology, availability of the rope elevator system has a high rank of status. System availability is a particular arrangement of reliability and maintainability. It is expected standard of the performance of the system under the specified conditions [4, 5]. For good functioning and high reliability of the elevators, all of its components and power supplies should be in adequate form and much more reliable, and power supply must be continuous [6]. The accessibility of the system usually refers to its maintenance, which is affected by adjusting the number of fault coverage and certain repair facility. Every organization wants high availability, so they prioritize maximum profit with less expenditure. Some authors, including Brunel [7], Gennheimer [8], Nelsen [9], Ram and Singh [10, 11], assumed more than one repair between two adjacent states, and the system is studied by the copula. The fault detection plays a significant role in the optimization of availability by controlling the maintenance cost when the system has partially failed. The models that determined the system dependability and incorporated fault coverage parameter are deliberated by Amari et al. [12], Dugan and Trivedi [13]. The fault coverage parameter is called coverage factor and defined as the probability of a number of faults recover to a number of faults occur [14, 15]. Therefore,

$$\text{Coverage factor} = \text{probability}[\text{system recovers}/\text{fault occurs}]$$

Some of the researchers have been done the work regarding elevators. Available literature reflects that the several approaches have been used to analyze the system functioning of elevators. But none of them think about the reliability and maintainability of the elevators, while reliability has a wide history in the field of engineering system. Many researchers, including Goel et al. [16], Gonzalez et al. [17], Gupta and Tewari [18], Ram and Goyal [19], Ram et al. [20], Yang and Dhillon [21], developed various stochastic models of the different industrial complex system and measured the various parameters using traditional reliability methods. Here, the authors developed the stochastic model of rope elevators under the consideration of several errors such as design error, maintenance error, equipment's fault for the investigation. These types of errors can have been prevented if safety instructions have been correctly followed and supported in the maintenance system. Under the consideration of these types of errors, four types of failure, such as hardware, software, human, and electricity failure, are taken account into the analysis of reliability indices of elevator. The generalized expressions of state transition probabilities have been derived using the process, supplementary variable technique, and Laplace transformation. The authors employed the fault coverage technique and Gumbel-Hougaard family of copula to enhance the life of the elevator.

2 Problem Descriptions

Modern elevators are divided into several sections, corresponding to the processes involved in the operation of elevators. In the beginning, elevators are in idle form and wait for the passengers. After loading passengers, it has two options, either goes up or down, after moving up, and it stops to its destination and starts unloading. After that it again has two options, either goes up or down and starts the same process again and again. In general, elevators consist of eight constituents, namely pulley, motor, gearbox, guide rail, counterweight, ropes, governor, software controller, and the functioning of these is taken account into the elevator's performance. So, the performance of elevators is affected by many types of failures that occur in the process as well as in hardware part.

The designed system can fail due to four types of failure, i.e., by the failure of hardware part, software error and human error, and due to electricity also. It is well known that without power supply, the system cannot survive because it operated through motor, and each component is attached to the power supply through the motor, so, we need continuous electricity and also a power backup on standby mode. The whole configuration of elevator system is shown in Fig. 1 with the help

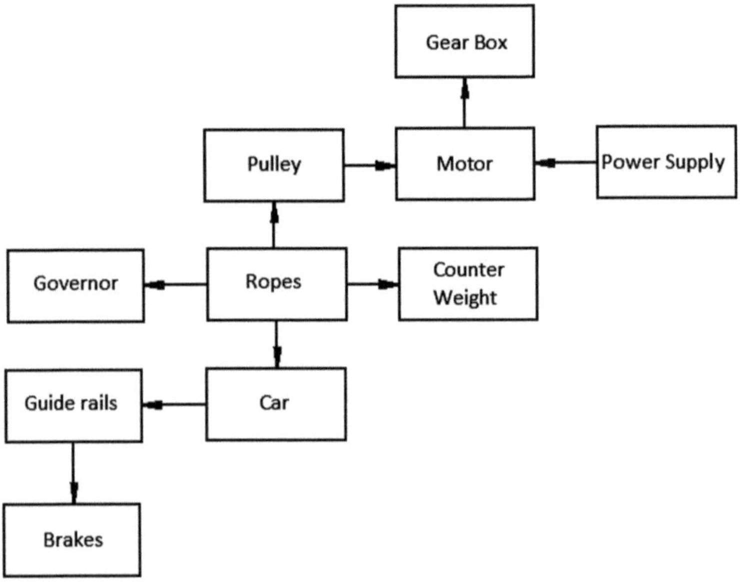

Fig. 1 Flowchart

of flowcharts. On the basis of this, draw a state transition diagram which is revealed in Fig. 2 and demonstrates all possible transitions of elevators under the consideration of each type of error. The system consists of three types of states good, degraded and failed. Each state of the designed system is described in Table 1.

The designed elevator system goes to failed state due to the failure of

- Motor;
- Both power supply and standby;
- Governor;
- Counterweight;
- Pulley;
- Guide rail;
- Human error.

The elevator system can work with less efficiency after some failure, i.e., the system goes to the degraded state after the failure of

- Ropes;
- Power supply;
- Gearbox;
- Software problem.

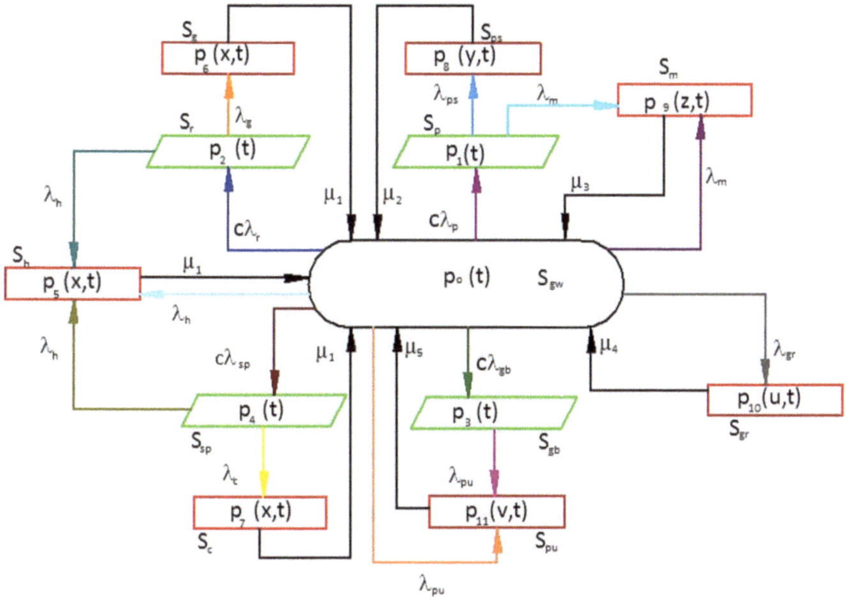

Fig. 2 State transition diagram

Table 1 State description

State	Description
S_{gw}	Good working state of the elevator
S_{p}	Degraded state of the system after the failure of power supply
S_{ps}	Complete failed state of the system due to failure of power supply and standby
S_{m}	State of elevator when motor has been failed
S_{r}	State of the system when rope or ropes have been failed
S_{g}	State of the elevator due to governor failure
S_{gr}	Complete failed state of the elevator system after the failure of guide rail
S_{gb}	Partially failed state of the system due to gearbox failure
S_{pu}	Due to failure of pulley, the system is in complete failed state
S_{sp}	Partially failed state of the system due to software problem
S_{c}	Complete failed state of the system due to the failure of counterweight
S_{h}	Complete failed state of the system due to human error

Table 2 Notations

t	Time scale
s	Laplace transform variable
c	Coverage factor
$\lambda_{m}/\lambda_{gr}/\lambda_{c}$	Failure rate of motor/guide rail/counterweight
λ_{p}/λ_{ps}	Failure rate of power supply and standby
$\lambda_{r}/\lambda_{g}/\lambda_{gb}$	Failure rate of ropes/governor/gearbox
λ_{pu}	Failure rate for the pulley
λ_{h}/λ_{sp}	Human/software error rate
μ_1	Constant repair rate from state $S_g/S_h/S_c$ to S_{gw}
μ_2	Constant repair rate from state S_{ps} to S_{gw}
μ_3	Constant repair rate for the system from state S_m to S_{gw}
μ_4	Constant repair rate for the elevator system from state S_{gr} to S_{gw}
μ_5	Constant repair rate for the elevator system from state S_{gr} to S_{gw}
$\bar{p}(s)$	Laplace transformation of $p(t)$
$p_i(t)$	The probability of good or partially failed state S_j of the system; i = 0, 1, 2, 3, 4; $j = gw, p, r, gb, sp$
$p_i(x,t)$	Probability density function of the complete failed state S_j of the system, at epoch time t and has an elapsed repair time x; i = 5, 6, 7; $j = h, g, c$
$p_{up}(t)$	Upstate system probability at time t or availability of the elevator system
$Rl(t)$	The reliability of the system at time t
$E_p(t)$	Expected profit during the interval [0, t)
K_1/K_2	Revenue/service cost per unit time, respectively
$u_1 = e^{\alpha}$, $u_2 = \mu_i(\alpha)$	The joint probability (failed state to good working state) according to Gumbel-Hougaard family is given as: $\exp\left[\alpha^{\theta} + \{\log \mu_i(\alpha)\}^{\theta}\right]^{1/\theta}$; i = 1, 2, 3, 4, 5; $\alpha = x, y, z, u, v$

3 Assumptions and Nomenclature

The following assumptions are used throughout the model, and notations are labeled in Table 2.

(i) Initially, the system is free from all types of failures.
(ii) The system may work with reduced capacity.
(iii) Faults can be covered perfectly from the degraded state.
(iv) Elevator system can repair after complete failure.
(v) Failure and repairs are constant.
(vi) At any time, sufficient repair facilities are available.
(vii) After repair, the efficiency of the system is not reduced.
(viii) The expression for the joint probability distribution of repair rates from the completed failed states is computed with the help of Gumbel-Hougaard family copula.

4 Formulation of the Model

By the probability of consideration and continuity of arguments, the following differential equations possess the present mathematical model

$$\left[\frac{\partial}{\partial t} + c\lambda_{\mathrm{p}} + c\lambda_{\mathrm{r}} + c\lambda_{\mathrm{gb}} + c\lambda_{\mathrm{sp}} + \lambda_{\mathrm{h}} + \lambda_{\mathrm{gr}} + \lambda_{\mathrm{m}} + \lambda_{\mathrm{pu}}\right] p_0(t)$$
$$= \int_0^\infty [p_5(x,t) + p_6(x,t) + p_7(x,t)]\mu_1(x)\mathrm{d}x$$
$$+ \int_0^\infty p_8(y,t)\mu_2(y)\mathrm{d}y + \int_0^\infty p_9(z,t)\mu_3(z)\mathrm{d}z \tag{1}$$
$$+ \int_0^\infty p_{10}(u,t)\mu_4(u)\mathrm{d}u + \int_0^\infty p_{11}(v,t)\mu_5(v)\mathrm{d}v$$

$$\left[\frac{\partial}{\partial t} + \lambda_{\mathrm{ps}} + \lambda_{\mathrm{m}}\right] p_1(t) = c\lambda_{\mathrm{p}} p_0(t) \tag{2}$$

$$\left[\frac{\partial}{\partial t} + \lambda_{\mathrm{g}} + \lambda_{\mathrm{h}}\right] p_2(t) = c\lambda_{\mathrm{r}} p_0(t) \tag{3}$$

$$\left[\frac{\partial}{\partial t} + \lambda_{\mathrm{pu}}\right] p_3(t) = c\lambda_{\mathrm{gb}} p_0(t) \tag{4}$$

$$\left[\frac{\partial}{\partial t} + \lambda_{\mathrm{c}} + \lambda_{\mathrm{h}}\right] p_4(t) = c\lambda_{\mathrm{sp}} p_0(t) \tag{5}$$

$$\left[\frac{\partial}{\partial t} + \frac{\partial}{\partial x} + \mu_1(x)\right] p_i(x,t) = 0; \quad i = 5,6,7 \tag{6}$$

$$\left[\frac{\partial}{\partial t} + \frac{\partial}{\partial y} + \mu_2(y)\right] p_8(y,t) = 0 \tag{7}$$

$$\left[\frac{\partial}{\partial t} + \frac{\partial}{\partial z} + \mu_3(z)\right] p_9(z,t) = 0 \tag{8}$$

$$\left[\frac{\partial}{\partial t} + \frac{\partial}{\partial u} + \mu_4(u)\right] p_{10}(u,t) = 0 \tag{9}$$

$$\left[\frac{\partial}{\partial t} + \frac{\partial}{\partial v} + \mu_5(v)\right] p_{11}(v,t) = 0 \tag{10}$$

Boundary conditions

$$p_5(0,t) = \lambda_h(p_0(t) + p_2(t) + p_4(t)) \tag{11}$$

$$p_6(0,t) = \lambda_{\mathrm{g}} p_2(t) \tag{12}$$

$$p_7(0,t) = \lambda_{\mathrm{c}} p_4(t) \tag{13}$$

$$p_8(0,t) = \lambda_{\mathrm{ps}} p_1(t) \tag{14}$$

$$p_9(0,t) = \lambda_{\mathrm{m}}[p_0(t) + p_1(t)] \tag{15}$$

$$p_{10}(0,t) = \lambda_{\mathrm{gr}} p_0(t) \tag{16}$$

$$p_{11}(0,t) = \lambda_{\mathrm{pu}}[p_0(t) + p_3(t)] \tag{17}$$

Initial condition

$P_0(0) = 1$ and other state probabilities are zero at $t = 0$.

5 Solution of the Model

Taking the Laplace transformation of Eqs. (1)–(17) using initial condition, we get

$$\begin{aligned}
&\left[s + c\lambda_{\mathrm{p}} + c\lambda_{\mathrm{r}} + c\lambda_{\mathrm{gb}} + c\lambda_{\mathrm{sp}} + \lambda_{\mathrm{h}} + \lambda_{\mathrm{gr}} + \lambda_{\mathrm{m}} + \lambda_{\mathrm{pu}}\right]\bar{p}_0(s) \\
&= 1 + \int_0^\infty [\bar{p}_5(x,s) + \bar{p}_6(x,s) + \bar{p}_7(x,s)]\mu_1(x)\mathrm{d}x \\
&+ \int_0^\infty \bar{p}_8(y,s)\mu_2(y)\mathrm{d}y + \int_0^\infty \bar{p}_9(z,s)\mu_3(z)\mathrm{d}z \\
&+ \int_0^\infty \bar{p}_{10}(u,s)\mu_4(u)\mathrm{d}u + \int_0^\infty \bar{p}_{11}(v,s)\mu_5(v)\mathrm{d}v
\end{aligned} \tag{18}$$

$$\left[s + \lambda_{\mathrm{ps}} + \lambda_{\mathrm{m}}\right]\bar{p}_1(s) = c\lambda_{\mathrm{p}}\bar{p}_0(s) \tag{19}$$

$$\left[s + \lambda_{\mathrm{g}} + \lambda_{\mathrm{h}}\right]\bar{p}_2(s) = c\lambda_{\mathrm{r}}\bar{p}_0(s) \tag{20}$$

$$\left[s + \lambda_{\mathrm{pu}}\right]\bar{p}_3(s) = c\lambda_{\mathrm{gb}}\bar{p}_0(s) \tag{21}$$

$$[s+\lambda_{c}+\lambda_{h}]\bar{p}_4(s)=c\lambda_{sp}\bar{p}_0(s) \tag{22}$$

$$\left[s+\frac{\partial}{\partial x}+\mu_1(x)\right]\bar{p}_i(x,s)=0;\quad i=5,6,7 \tag{23}$$

$$\left[s+\frac{\partial}{\partial y}+\mu_2(y)\right]\bar{p}_8(y,s)=0 \tag{24}$$

$$\left[s+\frac{\partial}{\partial z}+\mu_3(z)\right]\bar{p}_9(z,s)=0 \tag{25}$$

$$\left[s+\frac{\partial}{\partial u}+\mu_4(u)\right]\bar{p}_{10}(u,s)=0 \tag{26}$$

$$\left[s+\frac{\partial}{\partial v}+\mu_5(v)\right]\bar{p}_{11}(v,s)=0 \tag{27}$$

$$\bar{p}_5(0,s)=\lambda_{h}(\bar{p}_0(s)+\bar{p}_2(s)+\bar{p}_4(s)) \tag{28}$$

$$\bar{p}_6(0,s)=\lambda_{g}\bar{p}_2(s) \tag{29}$$

$$\bar{p}_7(0,s)=\lambda_{c}\bar{p}_4(s) \tag{30}$$

$$\bar{p}_8(0,s)=\lambda_{ps}\bar{p}_1(s) \tag{31}$$

$$\bar{p}_7(0,s)=\lambda_{m}[\bar{p}_0(s)+\bar{p}_1(s)] \tag{32}$$

$$\bar{p}_{10}(0,s)=\lambda_{gr}\bar{p}_0(s) \tag{33}$$

$$\bar{p}_{11}(0,s)=\lambda_{pu}[\bar{p}_0(s)+\bar{p}_3(s)] \tag{34}$$

6 State Transition Probability

Authors obtained state transition probabilities after solving Eqs. (18)–(34), as

$$\bar{p_0}(s)=\frac{1}{D(s)} \tag{35}$$

$$\bar{p}_1(s)=A_0(s)\bar{p}_0(s) \tag{36}$$

$$\bar{p}_2(s)=A_1(s)\bar{p}_0(s) \tag{37}$$

$$\bar{p}_3(s) = A_2(s)\bar{p}_0(s) \tag{38}$$

$$\bar{p}_4(s) = A_3(s)\bar{p}_0(s) \tag{39}$$

$$\bar{p}_5(s) = \lambda_{\mathrm{h}}\left(\frac{1-\overline{S}_{\mu_1}(s)}{s}\right)\left(1+\frac{c\lambda_{\mathrm{r}}}{s+\lambda_{\mathrm{g}}+\lambda_{\mathrm{h}}}+\frac{c\lambda_{\mathrm{sp}}}{s+\lambda_{\mathrm{c}}+\lambda_{\mathrm{h}}}\right)\bar{p}_0(s) \tag{40}$$

$$\bar{p}_6(s) = \left(\frac{1-\overline{S}_{\mu_1}(s)}{s}\right)\left(\frac{c\lambda_{\mathrm{r}}\lambda_{\mathrm{g}}}{s+\lambda_{\mathrm{g}}+\lambda_{\mathrm{h}}}\right)\bar{p}_0(s) \tag{41}$$

$$\bar{p}_7(s) = \left(\frac{1-\overline{S}_{\mu_1}(s)}{s}\right)\left(\frac{c\lambda_{\mathrm{sp}}\lambda_{\mathrm{c}}}{s+\lambda_{\mathrm{c}}+\lambda_{\mathrm{h}}}\right)\bar{p}_0(s) \tag{42}$$

$$\bar{p}_8(s) = \left(\frac{1-\overline{S}_{\mu_2}(s)}{s}\right)\left(\frac{c\lambda_{\mathrm{ps}}\lambda_p}{s+\lambda_{\mathrm{ps}}+\lambda_{\mathrm{m}}}\right)\bar{p}_0(s) \tag{43}$$

$$\bar{p}_9(s) = \lambda_{\mathrm{m}}\left(\frac{1-\overline{S}_{\mu_3}(s)}{s}\right)\left(1+\frac{c\lambda_{\mathrm{p}}}{s+\lambda_{\mathrm{ps}}+\lambda_{\mathrm{m}}}\right)\bar{p}_0(s) \tag{44}$$

$$\bar{p}_{10}(s) = \lambda_{\mathrm{gr}}\left(\frac{1-\overline{S}_{\mu_4}(s)}{s}\right)\bar{p}_0(s) \tag{45}$$

$$\bar{p}_{11}(s) = \lambda_{\mathrm{pu}}\left(\frac{1-\overline{S}_{\mu_5}(s)}{s}\right)\left(1+\frac{c\lambda_{\mathrm{gb}}}{s+\lambda_{\mathrm{pu}}}\right)\bar{p}_0(s) \tag{46}$$

where

$$\begin{aligned} D(s) = (s+B_1) - \big\{ & B_2 + B_3A_0(s) + A_1(s)\overline{S}_{\mu_1}(s)(\lambda_{\mathrm{h}}+\lambda_{\mathrm{g}}) \\ & + \lambda_{\mathrm{pu}}A_2(s)\overline{S}_{\mu_5}(s) + A_3(s)\overline{S}_{\mu_1}(s)(\lambda_{\mathrm{h}}+\lambda_{\mathrm{c}})\big\} \end{aligned}$$

$$\begin{aligned} B_1 &= c\lambda_{\mathrm{p}} + c\lambda_{\mathrm{r}} + c\lambda_{\mathrm{gb}} + c\lambda_{\mathrm{sp}} + \lambda_{\mathrm{h}} + \lambda_{\mathrm{gr}} + \lambda_{\mathrm{m}} + \lambda_{\mathrm{pu}}; \\ B_2 &= \lambda_{\mathrm{h}}\overline{S}_{\mu_1}(s) + \lambda_{\mathrm{gr}}\overline{S}_{\mu_4}(s) + \lambda_{\mathrm{m}}\overline{S}_{\mu_3}(s) + \lambda_{\mathrm{pu}}\overline{S}_{\mu_5}(s); \\ B_3 &= \lambda_{\mathrm{ps}}\overline{S}_{\mu_2}(s) + \lambda_{\mathrm{m}}\overline{S}_{\mu_3}(s) \end{aligned}$$

$$\begin{aligned} A_0(s) &= \frac{c\lambda_{\mathrm{p}}}{s+\lambda_{\mathrm{ps}}+\lambda_{\mathrm{m}}}; \quad A_1(s) = \frac{c\lambda_{\mathrm{r}}}{s+\lambda_{\mathrm{g}}+\lambda_{\mathrm{h}}}; \\ A_2(s) &= \frac{c\lambda_{\mathrm{gb}}}{s+\lambda_{\mathrm{pu}}}; \quad A_3(s) = \frac{c\lambda_{\mathrm{sp}}}{s+\lambda_{\mathrm{c}}+\lambda_{\mathrm{h}}}. \end{aligned}$$

7 Probability of the System in Upstate and Downstate

Through the predicted value of probability of all possible transition states, authors determine the probability of upstate and downstate system as

$$\begin{aligned}\overline{P}_{\text{up}}(s) &= \sum_{i=0}^{4} \overline{P}_i(s) \\ &= \left(1 + \sum_{i=0}^{3} A_i(s)\right)\bar{p}_0(s)\end{aligned} \tag{47}$$

$$\begin{aligned}\bar{P}_{\text{down}}(s) &= \sum_{i=5}^{11} \bar{P}_i(s) \\ &= \left\{\left(\frac{1-\bar{S}_{\mu_1}(s)}{s}\right)\left[\lambda_{\text{h}} + (\lambda_{\text{h}} + \lambda_{\text{g}})A_1(s) + (\lambda_{\text{h}} + \lambda_{\text{c}})A_3(s)\right] + \lambda_{\text{ps}}\left(\frac{1-\bar{S}_{\mu_2}(s)}{s}\right)A_0(s)\right. \\ &\quad \left. + \lambda_{\text{m}}\left(\frac{1-\bar{S}_{\mu_3}(s)}{s}\right)(1 + A_0(s)) + \lambda_{\text{gr}}\left(\frac{1-\bar{S}_{\mu_4}(s)}{s}\right) + \lambda_{\text{pu}}\left(\frac{1-\bar{S}_{\mu_5}(s)}{s}\right)(1 + A_2(s))\right\}\bar{p}_0(s)\end{aligned} \tag{48}$$

8 Particular Cases and Numerical Examples

8.1 Availability Analysis

Substituting the values of different failure rates as $\lambda_{\text{m}} = 0.007$, $\lambda_{\text{p}} = 0.326$, $\lambda_{\text{ps}} = 0.092$, $\lambda_{\text{g}} = 0.049$, $\lambda_{\text{r}} = 0.023$, $\lambda_{\text{h}} = 0.049$, $\lambda_{\text{sp}} = 0.233$, $\lambda_{\text{c}} = 0.049$, $\lambda_{\text{gb}} = 0.094$, $\lambda_{\text{pu}} = 0.092$, $\lambda_{\text{gr}} = 0.025$ [22] in Eq. (47) and taking the inverse Laplace transform, authors obtain the availability in three different situations by assuming the values of repair rates as

(i) When complete repair facility is available.
Setting the value of repair rates as $\mu_1 = 1$, $\mu_2 = 1$, $\mu_3 = 1$, $\mu_4 = 1$, $\mu_5 = 1$, one gets the availability of the elevator as shown in Table 3 and Fig. 3.

(ii) When incomplete repair facility is available.
Setting the value of repair rates as $\mu_1 = 0.5$, $\mu_2 = 0.6$, $\mu_3 = 0.7$, $\mu_4 = 0.75$, $\mu_5 = 0.65$, one obtains the availability as revealed in Table 4 and Fig. 4.

(iii) When repaired the elevator under Gumbel-Hougaard family of copula.
Setting the value of repair rates as $\mu_1(x) = 1$, $\mu_2(y) = 1$, $\mu_3(z) = 1$, $\mu_4(u) = 1$, $\mu_5(v) = 1$, $x = 1$, $y = 1$, $z = 1$, $u = 1$, $v = 1$, $\theta = 1$, one attains the availability as exposed in Table 5 and Fig. 5.
The comparative study of these three cases of elevator's availability is demonstrated by Figs. 6, 7, 8, 9, and 10.

Table 3 Availability with complete repair

Time t	Availability				
	c = 0.1	c = 0.3	c = 0.5	c = 0.7	c = 0.9
0	1.00000	1.00000	1.00000	1.00000	1.00000
1	0.89976	0.90274	0.90546	0.90793	0.91018
2	0.87098	0.87847	0.88462	0.88970	0.89391
3	0.86390	0.87508	0.88338	0.88959	0.89426
4	0.86323	0.87707	0.88635	0.89267	0.89704
5	0.86432	0.87994	0.88948	0.89544	0.89927
6	0.86575	0.88253	0.89191	0.89736	0.90068
7	0.86713	0.88460	0.89363	0.89859	0.90149
8	0.86835	0.88620	0.89482	0.89934	0.90195
9	0.86940	0.88743	0.89562	0.89980	0.90220
10	0.87031	0.88835	0.89616	0.90008	0.90234
11	0.87108	0.88905	0.89652	0.90024	0.90241
12	0.87174	0.88957	0.89677	0.90034	0.90245
13	0.87230	0.88997	0.89693	0.90040	0.90247
14	0.87278	0.89027	0.89704	0.90044	0.90249
15	0.87319	0.89049	0.89711	0.90046	0.90249
16	0.87354	0.89066	0.89716	0.90048	0.90250
17	0.87384	0.89079	0.89720	0.90048	0.90250
18	0.87409	0.89089	0.89722	0.90049	0.90250
19	0.87430	0.89096	0.89723	0.90049	0.90250
20	0.87449	0.89102	0.89725	0.90050	0.90251

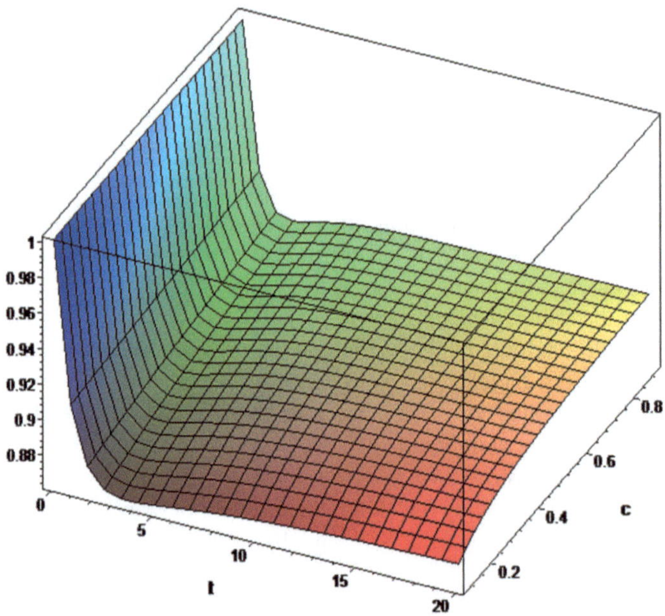

Fig. 3 Availability with complete repairs

Table 4 Availability with incomplete repairs

Time t	Availability				
	$c = 0.1$	$c = 0.3$	$c = 0.5$	$c = 0.7$	$c = 0.9$
0	1.00000	1.00000	1.00000	1.00000	1.00000
1	0.88238	0.88564	0.88845	0.89357	0.89381
2	0.83141	0.84003	0.84709	0.85412	0.85803
3	0.80987	0.82319	0.83314	0.84121	0.84649
4	0.80131	0.81812	0.82954	0.83770	0.84302
5	0.79845	0.81764	0.82955	0.83725	0.84210
6	0.79804	0.81878	0.83058	0.83764	0.84191
7	0.79862	0.82031	0.83173	0.83814	0.84190
8	0.79956	0.82177	0.83269	0.83854	0.84191
9	0.80058	0.82303	0.83343	0.83882	0.84192
10	0.80154	0.82405	0.83398	0.83901	0.84192
11	0.80242	0.82486	0.83436	0.83914	0.84192
12	0.80320	0.82548	0.83464	0.83922	0.84191
13	0.80388	0.82597	0.83482	0.83926	0.84191
14	0.80446	0.82635	0.83496	0.83930	0.84191
15	0.80496	0.82664	0.83505	0.83932	0.84191
16	0.80540	0.82686	0.83512	0.83934	0.84191
17	0.80577	0.82703	0.83516	0.83935	0.84192
18	0.80608	0.82716	0.83519	0.83936	0.84192
19	0.80636	0.82726	0.83522	0.83936	0.84192
20	0.80659	0.82733	0.83523	0.83936	0.84192

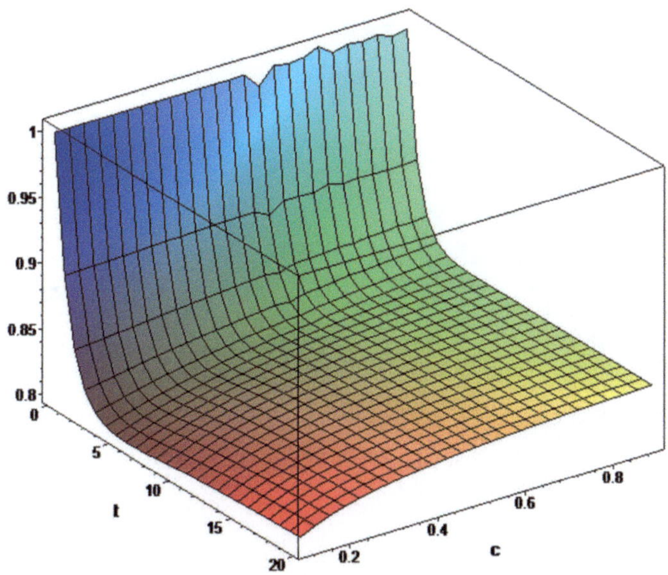

Fig. 4 Availability with incomplete repairs

Table 5 Availability under Gumbel-Hougaard family of copula repair

Time t	Availability with copula				
	$c = 0.1$	$c = 0.3$	$c = 0.5$	$c = 0.7$	$c = 0.9$
0	1.00000	1.00000	1.00000	1.00000	1.00000
1	0.94456	0.94654	0.94833	0.94994	0.95139
2	0.94277	0.94686	0.95015	0.95280	0.95493
3	0.94375	0.94932	0.95330	0.95616	0.95823
4	0.94473	0.95128	0.95547	0.95819	0.95998
5	0.94557	0.95274	0.95689	0.95936	0.96087
6	0.94628	0.95382	0.95782	0.96003	0.96132
7	0.94689	0.95464	0.95843	0.96041	0.96154
8	0.94740	0.95524	0.95882	0.96064	0.96166
9	0.94784	0.95569	0.95908	0.96076	0.96171
10	0.94821	0.95603	0.95926	0.96083	0.96174
11	0.94853	0.95628	0.95937	0.96088	0.96176
12	0.94879	0.95646	0.95944	0.96090	0.96177
13	0.94902	0.95660	0.95949	0.96092	0.96177
14	0.94922	0.95671	0.95952	0.96092	0.96177
15	0.94938	0.95678	0.95954	0.96093	0.96178
16	0.94952	0.95684	0.95955	0.96093	0.96178
17	0.94964	0.95689	0.95956	0.96094	0.96178
18	0.94974	0.95692	0.95957	0.96094	0.96178
19	0.94982	0.95694	0.95957	0.96094	0.96178
20	0.94990	0.95696	0.95958	0.96094	0.96178

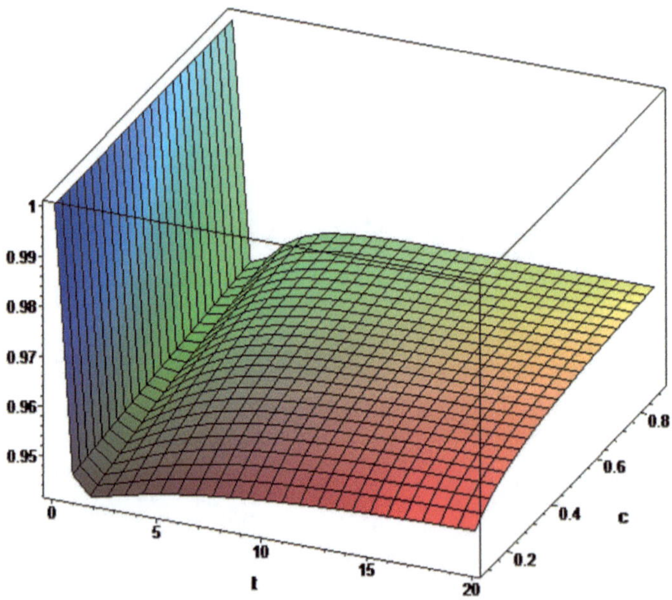

Fig. 5 Availability under Gumbel-Hougaard family of copula repair

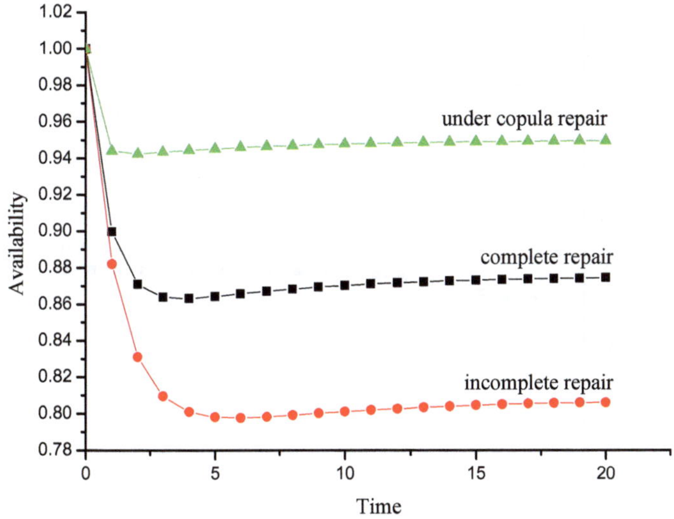

Fig. 6 Availability versus time at $c = 0.1$

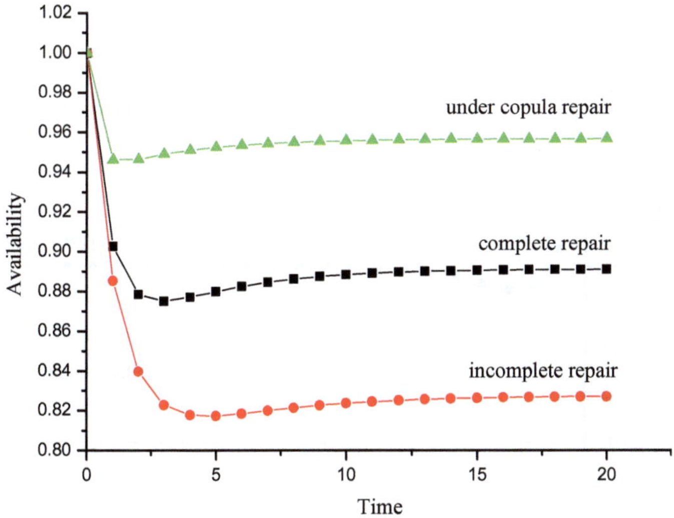

Fig. 7 Availability versus time at $c = 0.3$

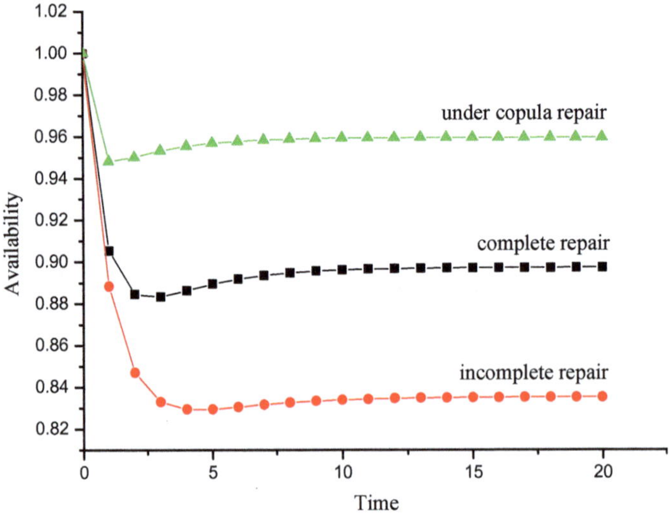

Fig. 8 Availability versus time at $c = 0.5$

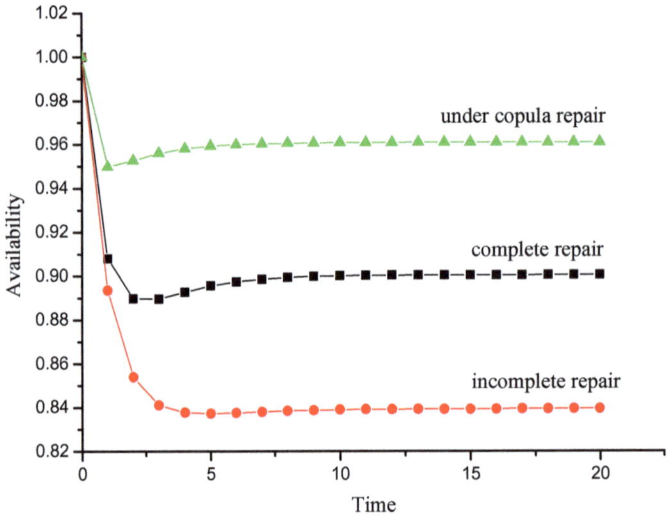

Fig. 9 Availability versus time at $c = 0.7$

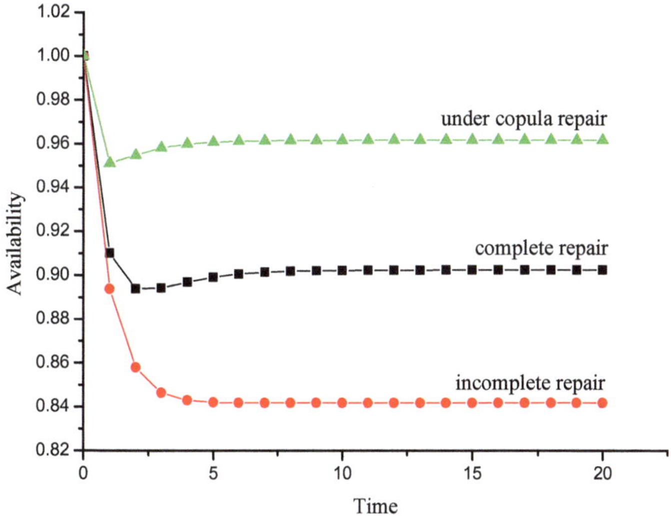

Fig. 10 Availability versus time at $c = 0.9$

8.2 *Reliability Analysis*

Considering the values of failure rates as $\lambda_m = 0.007$, $\lambda_p = 0.326$, $\lambda_{ps} = 0.092$, $\lambda_g = 0.049$, $\lambda_r = 0.023$, $\lambda_h = 0.049$, $\lambda_{sp} = 0.233$, $\lambda_c = 0.049$, $\lambda_{gb} = 0.094$, $\lambda_{pu} = 0.092$, $\lambda_{gr} = 0.025$ [22] and repair rates as zero in Eq. (47), and taking the inverse Laplace transformation. One can calculate the reliability of the elevator as shown in Table 6 and represented graphically in Fig. 11, after varying the time unit t from 0 to 20 and c = 0.1, 0.3, 0.5, 0.7, 0.9, respectively.

8.3 *Expected Profit*

Cost is the significant factor of all the financial assessments. It is convenient for development and cost prediction, evaluating the proficiency, measuring primacies, liability, and calculating impartiality [23, 24]. Expected profit in the elevator system depends upon the service or maintenance cost. Let us consider that the system has regular maintenance through always available service facility, then the expected profit during the interval (0, t] is given as

$$E_P(t) = K_1 \int_0^t P_{up}(t)\mathrm{d}t - tK_2 \tag{48}$$

Table 6 Reliability of the elevator

Time t	Reliability				
	$c = 0.1$	$c = 0.3$	$c = 0.5$	$c = 0.7$	$c = 0.9$
0	1.00000	1.00000	1.00000	1.00000	1.00000
1	0.84329	0.84731	0.85099	0.85437	0.85746
2	0.71477	0.72751	0.73824	0.74732	0.75503
3	0.60892	0.63169	0.64942	0.66335	0.67442
4	0.52133	0.55361	0.57692	0.59408	0.60694
5	0.44850	0.48884	0.51601	0.53485	0.54831
6	0.38766	0.43426	0.46366	0.48302	0.49632
7	0.33656	0.38760	0.41793	0.43701	0.44970
8	0.29344	0.34726	0.37749	0.39579	0.40767
9	0.25685	0.31201	0.34144	0.35869	0.36966
10	0.22565	0.28096	0.30913	0.32517	0.33524
11	0.19892	0.25344	0.28004	0.29485	0.30405
12	0.17591	0.22892	0.25380	0.26740	0.27577
13	0.15600	0.20698	0.23008	0.24251	0.25012
14	0.13870	0.18728	0.20862	0.21995	0.22687
15	0.12361	0.16956	0.18918	0.19950	0.20578
16	0.11039	0.15358	0.17156	0.18095	0.18664
17	0.09877	0.13916	0.15560	0.16412	0.16930
18	0.08852	0.12612	0.14112	0.14887	0.15356
19	0.07945	0.11433	0.12800	0.13503	0.13929
20	0.07141	0.10365	0.11610	0.12248	0.12634

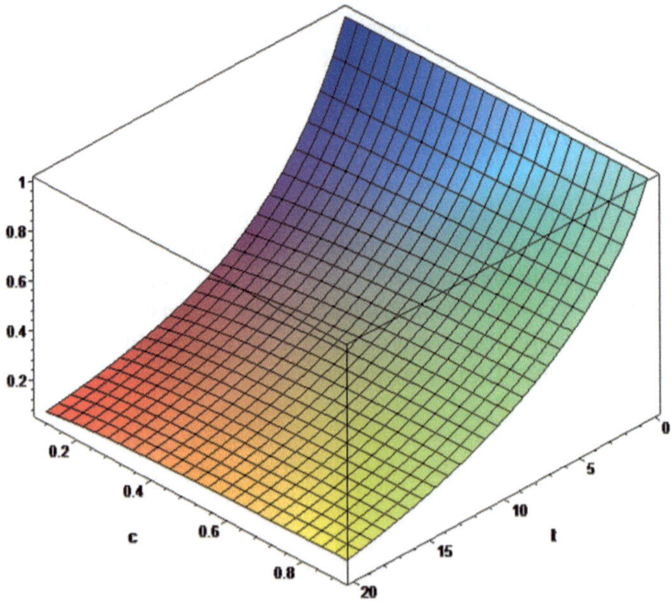

Fig. 11 Reliability versus time and coverage factor

Substituting the value in the inverse Laplace transformation of Eq. (46) in Eq. (48) and integrate, and fixing the value of failure rates as $\lambda_m = 0.007$, $\lambda_p = 0.326$, $\lambda_{ps} = 0.092$, $\lambda_g = 0.049$, $\lambda_r = 0.023$, $\lambda_h = 0.049$, $\lambda_{sp} = 0.233$, $\lambda_c = 0.049$, $\lambda_{gb} = 0.094$, $\lambda_{pu} = 0.092$, $\lambda_{gr} = 0.025$ [22]. One can find the expected profit in each case by considering the hundred percent fine revenue and varying the service cost as 0.1, 0.3, 0.5, respectively, and shown in Table 7, 8, and 9, and their graphical representation is demonstrated in Figs. 12, 13, and 14, respectively.

9 Results Interpretation

In this research work, the authors have studied the various reliability appearances such as accessibility, loyalty, and expected profit of the elevator system under the consideration of four types of failures and employing Markov process and supplementary variable technique. Through simulation and corresponding graphical representation, the authors made the following observations.

The behavior of availability of elevator system is studied in Figs. 3, 4, 5, 6, 7, 8, 9, and 10. The effects of coverage factor on elevator's availability in three different cases of repairs are revealed in Figs. 3, 4, and 5, and the graphs of elevator's availability shown in these figures expose that the availability of the elevator increases as coverage factor increases, while it decreases with the leading of time. The comparative study of elevator's availability under different cases of repair at a distinct value of coverage factor is represented in Figs. 6, 7, 8, 9, and 10. The graphs obtained in these figures show that the elevator system is highly available if it repairs under Gumbel-Hougaard family of copula, but, it is poorly available under distinct and incomplete repair at any value of coverage factor. With the complete repair, the designed elevator's system has availability lower than Gumbel-Hougaard family of copula and higher than the distinct and incomplete repair.

The trend of reliability of the designed elevator's system is demonstrated by a 3D graph in Fig. 11. The graph shows that the reliability of the elevator decreases as time increases, and it leads increment as coverage factor increases. Through the simulation and analysis of graph, one can see that reliability of the system varies between [1, 0]. At $t = 0$, system reliability is one and as t tends to infinity, reliability of the system approaches to zero.

The effects of fault coverage and service cost with respect to time on expected profit of elevator are optimized by the graphs epitomized in Figs. 12, 13, and 14. From the critical examination of these graphs, one can see that the expected profit in elevator system decreases as service or maintenance cost increases, and it leads increment with the passage of time. Expected profit of the elevator system also depends upon the number of fault coverage. The study of these graphs gives an idea that expected profit of the elevator also increases with the increment of coverage factor.

Table 7 Expected profit under complete repair

Time t	Expected profit														
	$c = 0.1$			$c = 0.3$			$c = 0.5$			$c = 0.7$			$c = 0.9$		
	$K_2 = 0.1$	$K_2 = 0.3$	$K_2 = 0.5$	$K_2 = 0.1$	$K_2 = 0.3$	$K_2 = 0.5$	$K_2 = 0.1$	$K_2 = 0.3$	$K_2 = 0.5$	$K_2 = 0.1$	$K_2 = 0.3$	$K_2 = 0.5$	$K_2 = 0.1$	$K_2 = 0.3$	$K_2 = 0.5$
0	0.00000	0.00000	0.00000	0.00000	0.00000	0.00000	0.00000	0.00000	0.00000	0.00000	0.00000	0.00000	0.00000	0.00000	0.00000
1	0.83994	0.63994	0.43994	0.84108	0.64108	0.44108	0.84214	0.64214	0.44214	0.84313	0.64313	0.44313	0.84406	0.64406	0.44406
2	1.62228	1.22228	0.82228	1.62868	1.22868	0.82868	1.63426	1.23426	0.83426	1.63915	1.23915	0.83915	1.64344	1.24344	0.84344
3	2.38880	1.78880	1.18880	2.40463	1.80463	1.20463	2.41756	1.81756	1.21756	2.42420	1.82420	1.22420	2.43703	1.83703	1.23703
4	3.15210	2.35210	1.55210	3.18053	2.38053	1.58053	3.20232	2.40232	1.60232	3.21928	2.41928	1.61928	3.23268	2.43268	1.63268
5	3.91582	2.91582	1.91582	3.95904	2.95904	1.95904	3.99028	2.99028	1.99028	4.01341	3.01341	2.01341	4.03091	3.03091	2.03091
6	4.68084	3.48084	2.28084	4.74031	3.54031	2.34031	4.78104	3.58104	2.38104	4.80987	3.60987	2.40987	4.83094	3.63094	2.43094
7	5.44729	4.04729	2.64729	5.52392	4.12392	2.72392	5.57386	4.17386	2.77386	5.60790	4.20790	2.80790	5.63207	4.23207	2.83207
8	6.21504	4.61504	3.01504	6.30937	4.70937	3.10937	6.36813	4.76813	3.16813	6.40689	4.80689	3.20689	6.43381	4.83381	3.23381
9	6.98394	5.18394	3.38394	7.09621	5.29621	3.49621	7.16338	5.36338	3.56338	7.20648	5.40648	3.60648	7.23590	5.43590	3.63590
10	7.75381	5.75381	3.75381	7.88413	5.88413	3.88413	7.95929	5.95929	3.95929	8.00643	6.00643	4.00643	8.03817	6.03817	4.03817
11	8.52452	6.32452	4.12452	8.67284	6.47284	4.27284	8.75564	6.55564	4.35564	8.80660	6.60660	4.40660	8.84055	6.64055	4.44055
12	9.29594	6.89594	4.49594	9.46216	7.06216	4.66216	9.55230	7.15230	4.75230	9.60690	7.20690	4.80690	9.64298	7.24298	4.84298
13	10.06798	7.46798	4.86798	10.25195	7.65195	5.05195	10.34915	7.74915	5.14915	10.40728	7.80728	5.20728	10.44545	7.84545	5.24545
14	10.84053	8.04053	5.24053	11.04207	8.24207	5.44207	11.14614	8.34614	5.54614	11.20770	8.40770	5.60770	11.24793	8.44793	5.64793
15	11.61352	8.61352	5.61352	11.83246	8.83246	5.83246	11.94322	8.94322	5.94322	12.00815	9.00815	6.00815	12.05042	9.05042	6.05042
16	12.38689	9.18689	5.98689	12.62304	9.42304	6.22304	12.74036	9.54036	6.34036	12.80862	9.60862	6.40862	12.85292	9.65292	6.45292
17	13.16058	9.76058	6.36058	13.41377	10.01377	6.61377	13.53754	10.13754	6.73754	13.60910	10.20910	6.80910	13.65542	10.25542	6.85542
18	13.93455	10.33455	6.73455	14.20462	10.60462	7.00462	14.33475	10.73475	7.13475	14.40959	10.80959	7.20959	14.45792	10.85792	7.25792
19	14.70875	10.90875	7.10875	14.99554	11.19554	7.39554	15.13198	11.33198	7.53198	15.21009	11.41009	7.61009	15.26042	11.46042	7.66042
20	15.48314	11.48314	7.48314	15.78654	11.78654	7.78654	15.92922	11.92922	7.92922	16.01058	12.01058	8.01058	16.06293	12.06293	8.06293

Table 8 Expected profit under incomplete repair

Time t	Expected profit														
	$c = 0.1$			$c = 0.3$			$c = 0.5$			$c = 0.7$			$c = 0.9$		
	$K_2 = 0.1$	$K_2 = 0.3$	$K_2 = 0.5$	$K_2 = 0.1$	$K_2 = 0.3$	$K_2 = 0.5$	$K_2 = 0.1$	$K_2 = 0.3$	$K_2 = 0.5$	$K_2 = 0.1$	$K_2 = 0.3$	$K_2 = 0.5$	$K_2 = 0.1$	$K_2 = 0.3$	$K_2 = 0.5$
0	0.00000	0.00000	0.00000	0.00000	0.00000	0.00000	0.00000	0.00000	0.00000	0.00000	0.00000	0.00000	0.00000	0.00000	0.00000
1	0.83315	0.63315	0.43315	0.83437	0.63437	0.43437	0.83538	0.63538	0.43538	0.83646	0.63646	0.43646	0.83806	0.63806	0.43806
2	1.58650	1.18650	0.78650	1.59364	1.19364	0.79364	1.59972	1.19972	0.79972	1.60522	1.20522	0.80522	1.61095	1.21095	0.81095
3	2.30556	1.70556	1.10556	2.32378	1.72378	1.12378	2.33852	1.73852	1.13852	2.35087	1.75087	1.15087	2.36222	1.76220	1.16222
4	3.01046	2.21046	1.41046	3.04384	2.24384	1.44384	3.06938	2.26938	1.46938	3.08953	2.28953	1.48953	3.10667	2.30668	1.50668
5	3.71004	2.71004	1.71004	3.76150	2.76150	1.76150	3.79878	2.79878	1.79878	3.82670	2.82670	1.82670	3.84916	2.84916	1.84916
6	4.40816	3.20816	2.00816	4.47965	3.27965	2.07965	4.52881	3.32881	2.12881	4.56403	3.36403	2.16402	4.59115	3.39115	2.19115
7	5.10644	3.70644	2.30644	5.19918	3.79918	2.39918	5.25998	3.85998	2.45998	5.30187	3.90187	2.50187	5.33306	3.93306	2.53306
8	5.80551	4.20551	2.60551	5.92024	4.32024	2.72024	5.99220	4.39221	2.79220	6.04019	4.44019	2.84019	6.07497	4.47497	2.87497
9	6.50558	4.70558	2.90558	6.64266	4.84266	3.04266	6.72529	4.92529	3.12529	6.77886	4.97886	3.17886	6.81689	5.01689	3.21689
10	7.20665	5.20665	3.20665	7.36622	5.36622	3.36622	7.45901	5.45901	3.45901	7.51778	5.51778	3.51778	7.55881	5.55881	3.55881
11	7.90864	5.70884	3.50864	8.09068	5.80968	3.69068	8.19319	5.99319	3.79319	8.25685	6.05685	3.85685	8.30073	6.10073	3.90073
12	8.61146	6.21146	3.81146	8.81587	6.41587	4.01587	8.92770	6.52770	4.12770	8.99603	6.59603	4.19603	9.04265	6.64265	4.24265
13	9.31501	6.71501	4.11501	9.54161	6.94161	4.34161	9.66244	7.06244	4.46244	9.73527	7.13527	4.53527	9.78456	7.18456	4.58456
14	10.01918	7.21918	4.41918	10.26778	7.46778	4.66778	10.39733	7.59733	4.79733	10.47455	7.67455	4.87455	10.52647	7.72647	4.92647
15	10.72390	7.72390	4.72390	10.99428	7.99428	4.99428	11.13234	8.13234	5.13234	11.21386	8.21386	5.21386	11.26839	8.26839	5.26839
16	11.42909	8.22909	5.02909	11.72103	8.52103	5.32103	11.86743	8.66743	5.46743	11.95319	8.75319	5.55319	12.01030	8.81030	5.67030
17	12.13468	8.73468	5.33468	12.44798	9.04798	5.64798	12.60257	9.20257	5.80257	12.69253	9.29253	5.89253	12.75223	9.35222	5.95222
18	12.84061	9.24061	5.64061	13.17507	9.57507	5.97507	13.33775	9.73775	6.13775	13.43188	9.83188	6.23188	13.49414	9.89414	6.29414
19	13.54683	9.74683	5.94683	13.90228	10.10228	6.30228	14.07295	10.27295	6.47295	14.17124	10.37124	6.57124	14.23606	10.43606	6.63606
20	14.25331	10.25331	6.25331	14.62958	10.62958	6.62958	14.80818	10.80818	6.80818	14.91060	10.91060	6.91060	14.97798	10.97798	6.97798

Table 9 Expected profit under copula repair

Time t	Expected profit														
	$c = 0.1$			$c = 0.3$			$c = 0.5$			$c = 0.7$			$c = 0.9$		
	$K_2 = 0.1$	$K_2 = 0.3$	$K_2 = 0.5$	$K_2 = 0.1$	$K_2 = 0.3$	$K_2 = 0.5$	$K_2 = 0.1$	$K_2 = 0.3$	$K_2 = 0.5$	$K_2 = 0.1$	$K_2 = 0.3$	$K_2 = 0.5$	$K_2 = 0.1$	$K_2 = 0.3$	$K_2 = 0.5$
0	0.00000	0.00000	0.00000	0.00000	0.00000	0.00000	0.00000	0.00000	0.00000	0.00000	0.00000	0.00000	0.00000	0.00000	0.00000
1	0.86015	0.66015	0.46015	0.86098	0.66098	0.46098	0.86174	0.66174	0.46174	0.86246	0.66246	0.46246	0.86313	0.66313	0.46313
2	1.70316	1.30316	0.90316	1.70708	1.30708	0.90708	1.71047	1.31047	0.91047	1.71340	1.31340	0.91340	1.71596	1.31596	0.91596
3	2.54640	1.94640	1.34640	2.55520	1.95520	1.35520	2.56228	1.96228	1.36228	2.56801	1.96801	1.36801	2.57270	1.97270	1.37270
4	3.39065	2.59065	1.79065	3.40555	2.60555	1.80555	3.41674	2.61674	1.81674	3.42528	2.62528	1.82528	3.43190	2.63190	1.83190
5	4.23581	3.23581	2.23581	4.25759	3.25759	2.25759	4.27296	3.27296	2.27296	4.28410	3.28410	2.28410	4.29238	3.29238	2.29238
6	5.08175	3.88175	2.68175	5.11090	3.91090	2.71090	5.13035	3.93035	2.73035	5.14383	3.94383	2.74383	5.15349	3.95349	2.75349
7	5.92835	4.52835	3.12835	5.96515	4.56515	3.16515	5.98850	4.58850	3.18850	6.00406	4.60406	3.20406	6.01493	4.61493	3.21493
8	6.77550	5.17550	3.57550	6.82010	5.22010	3.62010	6.84714	5.24714	3.64714	6.86460	5.26460	3.66460	6.87654	5.27654	3.67654
9	7.62313	5.82313	4.02313	7.67558	5.87558	4.07558	7.70610	5.90610	4.10610	7.72530	5.92530	4.12530	7.73823	5.93823	4.13823
10	8.47116	6.47116	4.47116	8.53145	6.53145	4.53145	8.56528	6.56528	4.56528	8.58611	6.58611	4.58611	8.59996	6.59996	4.59996
11	9.31953	7.11953	4.91953	9.38761	7.18761	4.98761	9.42460	7.22460	5.02460	9.44696	7.24696	5.04696	9.46171	7.26171	5.06171
12	10.16820	7.76820	5.36820	10.24398	7.84398	5.44398	10.28400	7.88400	5.48400	10.30786	7.90786	5.50786	10.32348	7.92348	5.52348
13	11.01711	8.41711	5.81711	11.10052	8.50052	5.90052	11.14347	8.54347	5.94347	11.16877	8.56877	5.96877	11.18525	8.58525	5.98525
14	11.86623	9.06623	6.26623	11.95718	9.15718	6.35718	12.00297	9.20297	6.40297	12.02969	9.22969	6.42969	12.04702	9.24702	6.44702
15	12.71553	9.71553	6.71553	12.81393	9.81393	6.81393	12.86250	9.86250	6.86250	12.89062	9.89062	6.89062	12.90879	9.90879	6.90879
16	13.56498	10.36498	7.16498	13.67075	10.47075	7.27075	13.72205	10.52205	7.32205	13.75155	10.55155	7.35155	13.77058	10.57058	7.37058
17	14.41456	11.01456	7.61456	14.52761	11.12761	7.72761	14.58161	11.18161	7.78161	14.61249	11.21249	7.81249	14.63235	11.23235	7.83235
18	15.26425	11.66425	8.06425	15.38452	11.78452	8.18452	15.44118	11.84118	8.24118	15.47342	11.87342	8.27342	15.49412	11.89412	8.29412
19	16.11403	12.31403	8.51403	16.24145	12.44145	8.64145	16.30075	12.50075	8.70075	16.33436	12.33436	8.73436	16.35590	12.55590	8.75590
20	16.96389	12.96389	8.96389	17.09840	13.09840	9.09840	17.16033	13.16033	9.16033	17.19530	13.19530	9.19530	17.21768	13.21768	9.21768

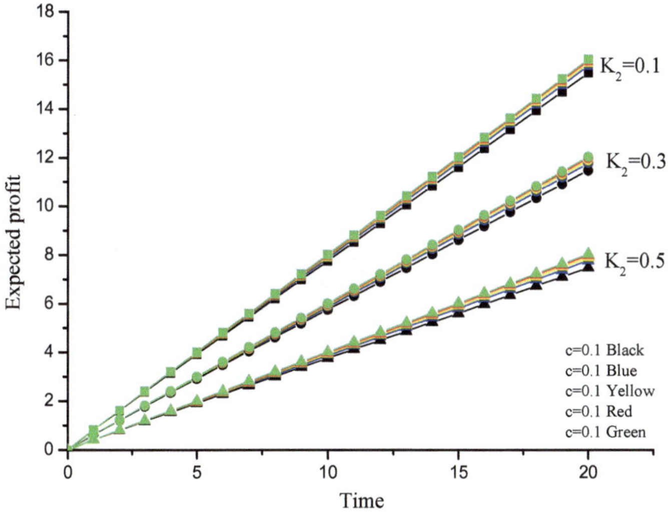

Fig. 12 Expected profit under complete repair

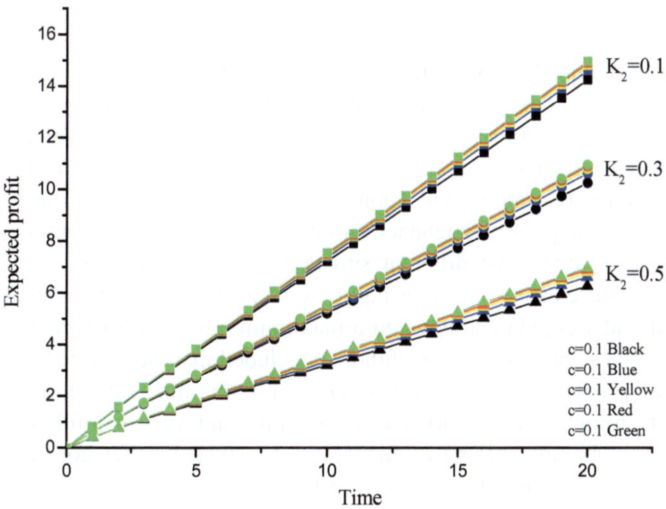

Fig. 13 Expected profit under incomplete repair

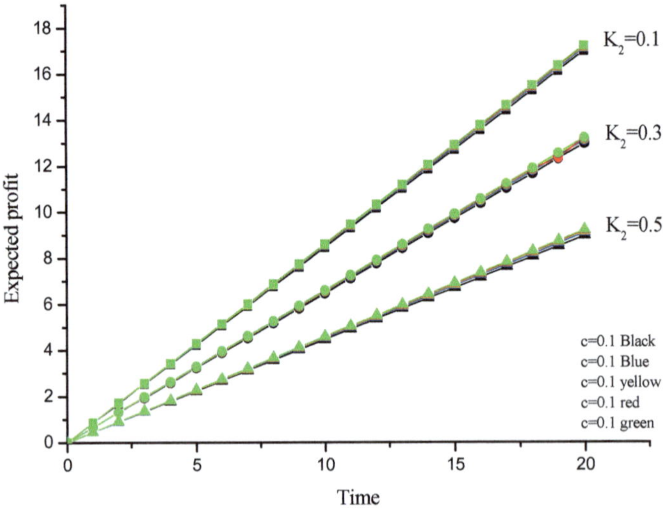

Fig. 14 Expected profit under Gumbel-Hougaard family of copula repair

10 Conclusion

This chapter intended to examine the reliability indices of rope elevator and try to find out some methodologies that enhance its performance. Maintenance and reliability are the key factor of elevator's performance. A number of safety challenges can be successfully fixed for the elevators to continue to make further improvements in reliability, safety, and economics. Many incidents in the elevator are generally based on poor maintenance and material used. So, the authors considered four types of errors, three different situations of repair and fault coverage. It is observed that the elevator system is highly reliable and has longtime availability if it is repaired under copula and covered a maximum number of detected faults. After the study of these parameters, the authors concluded that the elevator system attains high profit by maximizing the fault coverage and minimized the maintenance cost. The results found in this work are worthwhile and very helpful for engineers, designers, and manufacturers to achieve a highly available elevator system and for the regular maintenance of the system in future.

References

1. Poon OL (1993) U.S. Patent No. 5,230,406. U.S. Patent and Trademark Office, Washington, DC
2. Edoardo GB (1963) U.S. Patent No. 3,101,130. U.S. Patent and Trademark Office, Washington, DC
3. Koppensteiner W (1985) U.S. Patent No. 4,556,155. U.S. Patent and Trademark Office, Washington, DC

4. Avižienis A, Laprie JC, Randell B, Landwehr C (2004) Basic concepts and taxonomy of dependable and secure computing. IEEE Trans Dependable Secure Comput 1(1):11–33
5. Ram M (2013) On system reliability approaches: a brief survey. Int J Syst Assur Eng Manage 4(2):101–117
6. Janovský L (1999) Elevator mechanical design. Elevator World Inc., USA
7. Brunel NB, Lapuyade-Lahorgue J, Pieczynski W (2010) Modeling and unsupervised classification of multivariate hidden Markov chains with copulas. Autom Control IEEE Trans 55(2):338–349
8. Gennheimer H (2002) Model risk in copula based default pricing models, working paper series. Swiss Banking Inst., University of Zurich and NCCR FINRISK, Zurich, Switzlerland, Working paper no. 19
9. Nelsen RB (2013) An introduction to copulas, vol 139. Springer Science and Business Media, Berlin
10. Ram M, Singh SB (2008) Availability and cost analysis of a parallel redundant complex system with two types of failure under preemptive-resume repair discipline using Gumbel-Hougaard family copula in repair. Int J Reliab Qual Saf Eng 15(04):341–365
11. Ram M, Singh SB (2009) Analysis of reliability characteristics of a complex engineering system under copula. J Reliab Stat Stud 2(1):91–102
12. Amari SV, Dugan JB, Misra RB (1999) A separable method for incorporating imperfect fault-coverage into combinatorial models. IEEE Trans Reliab 48(3):267–274
13. Dugan JB, Trivedi KS (1989) Coverage modeling for dependability analysis of fault-tolerant systems. IEEE Trans Comput 38(6):775–787
14. Arnold TF (1973) The concept of coverage and its effect on the reliability model of a repairable system. IEEE Trans Comput 100(3):251–254
15. Bouricius WG, Carter WC, Schneider PR (1969). Reliability modeling techniques for self-repairing computer systems. In: Proceedings of the 24th national conference on ACM, August, pp 295–309
16. Goel LR, Sharma GC, Gupta P (1985) Stochastic analysis of a man-machine system with critical human error. Microelectron Reliab 25(4):669–674
17. González V, Gómez JF, López M, Crespo A, de León PM (2009) Availability and reliability assessment of industrial complex systems: a practical view applied on a bioethanol plant simulation. In: Martorell et al (eds) Safety, reliability and risk analysis: theory, methods and applications, pp 687–695
18. Gupta S, Tewari PC (2011) Performance modeling of power generation system of a thermal plant. Int J Eng 24(3):239–248
19. Ram M, Goyal N (2015) Gas turbine assimilation under copula-coverage approaches. In: Research advances in industrial engineering. Springer International Publishing, Berlin, pp 103–116
20. Ram M, Singh SB, Singh VV (2013) Stochastic analysis of a standby system with waiting repair strategy. Syst Man Cybern: Syst IEEE Trans 43(3):698–707
21. Yang N, Dhillon BS (1995) Availability analysis of a repairable standby human-machine system. Microelectron Reliab 35(11):1401–1413
22. Park ST, Yang BS (2010) An implementation of risk-based inspection for elevator maintenance. J Mech Sci Technol 24(12):2367–2376
23. Ram M, Manglik M (2014) Stochastic behaviour analysis of a Markov model under multi-state failures. Int J Syst Assur Eng Manage 5(4):686–699
24. Singh VV, Ram M, Rawal DK (2013) Cost analysis of an engineering system involving subsystems in series configuration. Autom Sci Eng, IEEE Trans 10(4):1124–1130

Index

M. Ram and J.P. Davim (eds.), *Modeling and Simulation in Industrial Engineering*, Management and Industrial Engineering, DOI 10.1007/978-3-319-60432-9

MIX
Papier aus verantwortungsvollen Quellen
Paper from responsible sources
FSC® C105338

If you have any concerns about our products,
you can contact us on
ProductSafety@springernature.com

In case Publisher is established outside the EU,
the EU authorized representative is:
Springer Nature Customer Service Center GmbH
Europaplatz 3, 69115 Heidelberg, Germany

Printed by Libri Plureos GmbH
in Hamburg, Germany